IAEI EXAM PREPARATION

PLAN REVIEW STUDY GUIDE

ELECTRICAL INSPECTOR CERTIFICATION PROGRAM

IAEI

The Electrical Safety Leader

The Electrical Safety Leader

International Association of Electrical Inspectors
901 Waterfall Way, Suite 602
Richardson, TX 75080-7702

 First edition published 1981
Printed in the United States of America
26 25 24 23 22 5 4 3 2 1

ISBN: 978-1-890659-90-5

Technical Editor: Mark Ode
Technical Review: Joseph Wages, Jr.
Production Technician: Laura Hildreth
Cover Design: Laura Hildreth

Notice to the Reader

This book has not been processed in accordance with NFPA Regulations Governing Committee Projects. Therefore, the text and commentary in it shall not be considered the official position of the NFPA or any of its committees and shall not be considered to be, nor relied upon as a formal interpretation of the meaning or intent of any specific provision or provisions of the 2020 edition of NFPA 70, *National Electrical Code*.®[1]

Publishers do not warrant or guarantee any of the products described herein or perform any independent analysis in connection with any of the product information contained herein. Publisher does not assume, and expressly disclaims, any obligation to obtain and include information referenced in this work.

The reader is expressly warned to consider carefully and adopt all safety precautions that might be indicated by the activities described herein and to avoid all potential hazards. By following the instructions contained herein, the reader willingly assumes all risks in connection with such instructions.

THE PUBLISHERS MAKE NO REPRESENTATIONS OR WARRANTIES OF ANY KIND, INCLUDING, BUT NOT LIMITED TO, THE IMPLIED WARRANTIES OF FITNESS FOR PARTICULAR PURPOSE, MERCHANTABILITY OR NON-INFRINGEMENT, NOR ARE ANY SUCH REPRESENTATIONS IMPLIED WITH RESPECT TO SUCH MATERIAL. THE PUBLISHERS SHALL NOT BE LIABLE FOR ANY SPECIAL, INCIDENTAL, CONSEQUENTIAL OR EXEMPLARY DAMAGES RESULTING, IN WHOLE OR IN PART, FROM THE READER'S USES OF OR RELIANCE UPON THIS MATERIAL.

TABLE OF CONTENTS

TECHNIQUES FOR TESTING

People study for a variety of reasons. Some study to improve their understanding and grasp of their chosen field, but the vast majority study because they will be required to show their competence in a given subject. Competence is generally measured by test scores. Unfamiliarity with the specific subject is one of the main reasons for low scores. Another reason is that a person's performance is only a sample of their behavior at that point in time. In other words, test scores are not set in concrete, and you can improve your score by preparing better, improving your attitude, and understanding the best methods to use in test taking.

The easiest way to improve your test performance is to reduce your anxiety level. The study guides sponsored by the International Association of Electrical Inspectors are designed to improve your understanding of how to find information in the *National Electrical Code*. The questions are designed to be similar to test formats you are likely to find in electrical inspector certification examinations. It should be understood that no effort is made to duplicate those questions exactly. To do so, would be nonproductive as questions are changed from time to time. Rather, the approach used in these study guides is to help you master locating information quickly and accurately. You can then master any test question. The refresher courses are available in three modules: *Electrical General Study Guide, One- and Two-Family Dwellings Study Guide*, and *Plan Review Study Guide*.

The goal of these courses is to ensure your best possible performance by understanding how to use the *National Electrical Code*® and thus reduce your anxiety level. The courses may be used for individual home study and are also ideally suited for an instructor/student classroom approach.

EXAMINATION FOR CERTIFICATION

Methods to Improve Your Test Scores

1. *Prepare by studying.* The questions in this refresher course are designed to teach you how to find the rules quickly and efficiently. If you learn how to use the table of contents and index properly, you will find the material in a reasonably short period. This knowledge and confidence will help reduce your anxiety and raise your scores. Consistent study leads to a higher retention level than cramming just prior to the examination date. The ultimate goal is to become familiar with the use of the *NEC* and testing format.

2. *Arrive on time.* Make sure you allow plenty of time to arrive promptly, but not too early. Since nervousness is contagious, don't associate too closely with the other examinees as you may pick up their anxiety level.

3. *Eliminate wrong answers.* If the test is multiple choice and you have four alternatives, the odds are 4 to 1 that you can guess the right answer. If you eliminate any two alternatives, your chances are increased to 50-50.

4. *Read directions carefully.* Many mistakes are made merely because the directions have been misunderstood. If, after reading carefully, you still are not sure, ask the proctor for clarification.

5 *Allow yourself enough time.* Based on the number of questions you have to answer, allot a specific amount of time for each question.

6 *Answer question first.* If the question has several alternatives, attempt to answer the question before you check the alternatives. In this way, you can evaluate your answer against the alternatives.

7 *Skip difficult questions.* If you are unsure of the answer or know that you are familiar with the material but don't have a ready answer, skip the question and go on to other material. Complete known questions then return to those you do not know. Chances are that your mind will subconsciously work out the answer so that it will be easier when you return to it.

8 *First choice is usually best.* If you pick an answer to a multiple-choice question and have later reservations about the right answer, remember that your first choice is usually best. If on later evaluation you know you have made a mistake, by all means, change your answer.

9 *Read questions carefully.* Make sure you note keywords that might change the meaning of the questions. Note negative disclaimers such as, "which of the following are not...". A handy way to increase understanding is to underline keywords. This action tends to channel your thinking along the right path.

10 *Make sure you are comfortable.* If you have on too many clothes, remove some. Being too warm tends to make you drowsy, which leads to a loss of concentration.

11 *Re-check your work.* The last thing you should do before handing in your paper is to recheck and make sure you have not made any clerical mistakes. Many times, you will know the right answer, but lose points on your score because of a clerical error.

Study Plan

Familiarize yourself completely with the codebook. Until one is familiar with the *Code* to the extent that location of specific requirements is committed to memory, use of the index is the best way to find information. The index contains an alphabetical list of what is in the *Code*, telling you where to find the topics covered.

Even though you may know the answer to a question, follow this sequence to establish the answer:

1 Check the table of contents to find the proper code article.

2 Select key words from the question that will identify the code article and subject matter that will be used to find the requirements in the index.

For example, you have been asked to verify the size of a grounding electrode conductor for a 200 ampere AC service supplying a dwelling. The dwelling unit has a metal underground water pipe, and the service-entrance conductors are 3/0, THWN, copper.

From the question, you can identify the subject of the question is a grounding electrode conductor. Specifically, you are being asked to determine the size. Additional keywords or subject matter are the *size of the service-entrance conductors* and the fact the dwelling is supplied by an *metal underground water pipe.*

Looking in the index, you find "Grounding electrode conductors" under which you will find "Sizing 250.30(A)(6)(a), 250.66 and 250.166." Scanning 250.30(A)(6) you quickly realize this is related to separately derived systems; so, the next choice was 250.66 from the index. Section 250.66 is titled "Size of Alternating-Current Grounding Electrode Conductors."

The opening paragraph of 250.66 states the grounding electrode conductor cannot be less than given in Table 250.66 except as permitted in (A)–(C). Scanning 250.66(A)–(C), you find they are not applicable to the question; so, Table 250.66 must be used.

Table 250.66 is based on the size of the ungrounded conductors; so, in our case, we determine a 4 AWG copper grounding electrode conductor is required for an AC service supplied by 3/0 copper service entrance conductors.

3 If, in the index, you do not readily find the location of the requirements related to the question, scan the boldface titles of the appropriate sections in the body of the code to locate quickly the subject material.

The Table of Contents lists, in numerical sequence, the subjects covered by each chapter and each article. So it provides the article number, part number if applicable, and a page number that can be used to find a location to start scanning section and subsection titles. Experienced users of the *NEC* are generally familiar with the content of *NEC* articles and, therefore, often use the Table of Contents to find a page number as a starting point.

4 Using the above question, an experienced user of the *NEC* would know that grounding and bonding requirements are found in Article 250. Scanning the information in the *NEC* Table of Contents, we find the requirements for the "Grounding Electrode System and Grounding Electrode Conductor" are in Part III of Article 250, and they start on page 120. Scanning the boldfaced section and subsection titles starting on page 120, we find 250.66 "Size of Alternating-Current Grounding Electrode Conductors on page 123."

This study guide is divided into subject categories, and each category has several questions. Each question is followed by the procedure for finding the correct answer. Follow the procedures step by step to learn good work habits. The answers are listed in the back of the book. Do not look up the answers until you have completed work in the entire category.

Remember, you are only shortchanging yourself by not following the step-by-step method of problem solving. The goal is to learn how to find code information in the most efficient manner.

This study guide is based on the *National Electrical Code*, 2020 edition.

HOW TO USE THE *NEC*

The *National Electrical Code*, sponsored by the National Fire Protection Association, is the most widely adopted code in the world. It is also, in all probability, the most widely misinterpreted code in the world. The *NEC* is adopted by federal, state and local governments as well as by private industry. As far as NFPA is concerned, the *NEC* is purely advisory. It becomes enforceable as law only upon adoption by an agency having authority to enforce its rules. Requirements for electrical inspection of installations, licensing of electrical contractors and electricians, as well as qualifications of electrical inspectors usually are contained in laws or ordinances that are associated with adoption and enforcement of electrical codes.

SCOPE OF THE CODE

Article 90 serves as an introduction to the *NEC* and provides a scope to specify the electrical installations that are covered and those that are not covered. The *Code* covers installations and removal of electric conductors and equipment within or on public and private buildings or other structures, including mobile homes, recreational vehicles, and floating buildings; and other premise wiring such as yards, carnival, parking, and other lots, and industrial substations. The *Code* also covers installations of conductors and equipment that connect to the supply of electricity, installations of other outside conductors and equipment on the premises, as well as installations supplying shore power to ships and watercraft in marinas and boat yards, those used to export electric power from vehicles to premises wiring, signaling and communications conductors and equipment and optical fiber cable.

In addition, the *Code* covers installations in buildings used by the electric utility, such as office buildings, warehouses, garages, machine shops and recreational buildings that are not an integral part of a generating plant, substation or control center.

The *Code* does not cover installations in ships, watercraft other than floating buildings, railway rolling stock, aircraft, or automotive vehicles other than mobile homes and recreational vehicles.

It also does not cover underground mines, railway conductors, or installations of communications equipment under the exclusive control of the communications utility. This latter requirement generally applies to overhead and underground communications conductors up to their termination in a locked room under the exclusive control of the communications utility. Installations, such as telephone distribution conductors, in hung ceilings or other accessible locations, are not normally under the exclusive control of the telephone company and are, therefore, covered by Articles 800 and 805. Electric utility generation, transmission, transformation, energy storage and distribution conductors including associated lighting are not normally covered by the *NEC*.

ENFORCEMENT

It is intended that the authority having jurisdiction interpret the rules in the *Code* and approve all devices, materials, equipment and conductors.

While the code is primarily intended to apply to new construction, the second paragraph of 90.4 permits the authority having jurisdiction to use his or her judgment in applying the code to rewiring in old installations.

By special permission, the authority having jurisdiction may waive specific requirements in this code or permit alternative methods where it is assured that equivalent objectives can be achieved by establishing and maintaining effective safety.

In discharging the responsibilities granted in 90.4, authorities having jurisdiction should enforce the requirements of 110.3(B) to assure that equipment is used as intended by the manufacturer and electrical products testing laboratory.

All mandatory rules are characterized by the use of the word "shall." Where the word "may" is used it means that the authority having jurisdiction has the prerogative of granting permission.

This should not be confused with the permissive term "may," which has largely been replaced in current editions of the code by the term "shall be permitted."

Informational notes are sometimes mistakenly enforced as mandatory requirements. They are intended to provide information or explanatory material and are not intended to be enforced as a part of the requirements. Where more than one Informational Note follows a section, they are numbered consecutively such as Informational Note No. 1, Informational Note No. 2, etc.

USE THE TABLE OF CONTENTS AND INDEX

To save time, one should use the table of contents to his or her best advantage. The table of contents lists in numerical sequence the subjects covered by each chapter and each article. Review of the table of contents gives an excellent overview of where the *Code* covers specific subjects or equipment.

Until one is familiar with the *Code* to the extent that location of specific requirements is committed to memory, use of the index is the best way to find information. The index contains in alphabetical order a list of what is in the *Code*, telling where to find topics covered. In some cases, the location of topics can be found in more than one way. For instance, the ampacities for conductors can be found under the heading "Ampacities" and also under the heading "Conductors."

Seemingly vague requirements can often become crystal clear by making use of the definitions in Article 100. For instance, the definition of *overload* says that a fault, such as a short circuit or ground fault, is not an overload.

CODE ARRANGEMENT

The arrangement of the *Code* is explained in 90.3. One should become thoroughly familiar with this concept for proper application of code rules. General rules in early chapters are sometimes supplemented or modified by later chapters.

Chapter Arrangement

Many articles have a scope that explains what is covered by that article. Mistakes in understanding and application of *Code* requirements are often avoided by reviewing the article scope before assuming general coverage of a requirement.

Chapter 8 covers communication systems and is not subject to the requirements of Chapters 1 through 7, except where the requirements are specifically referenced therein. For example, refer to 800.44(A)(3), Climbing Space. "The climbing space through communications wires and cables shall comply with the requirements of 225.14(D)." Without a reference like this, there would not be any requirements for climbing space relative to communications conductors unless specifically included in Article 800.

Numbering System

Essential to the understanding of any code or standard is familiarizing oneself with the arrangement and numbering system employed. The numbering system utilized for the *NEC* consists of the introduction Article 90, along with chapters numbered from 1 through 9. Chapters are further divided into several articles consisting of numbers in hundreds such as Article 100, Article 200, Article 300, etc. All articles in Chapter 1 begin with 100, Chapter 2 with 200, etc.

Some articles have several parts which are in roman numerical sequence such as I, II, III, IV, etc. Common requirements are grouped in parts of articles. Part I, General, usually contains provisions that apply to all the other parts of the article. Other parts contain provisions that are independent of all other parts; except Part I General. For instance, Part II of Article 680 contains provisions for permanently installed pools, and Part III contains provisions for storable pools. Each part is separate and independent of the other, but all parts must comply with the applicable general requirements of Part I; unless specifically stated otherwise.

All articles are divided into sections such as 110.3, 110.5, 110.8, 110.26, etc. Subsections are further numbered with upper case alphabetical characters similar to 110.26(A), 110.26(B), 110.26(C). In some cases, paragraphs following subsections are numbered similar to 110.27(A)(1), 110.27(A)(2), 110.27(A)(3), etc.

Parallel Numbering

The *NEC*-2002 introduced a new parallel numbering system. Common types of rules or information are assigned the same numbered section in each article. As an example, article scopes are designed as ".1" and definitions for the specific article are assigned as ".2". Uses permitted are ".10" in all articles. This will help users of the *Code* locate this type of common material without having to search for it in different locations in each article.

GENERAL REQUIREMENTS

Users often miss important requirements by failing to review the general requirements before deciding that the subject is not covered in the *Code*. Article 110 contains general requirements for electrical installations that are applied throughout all the other articles, unless modified by other articles. Article 300 provides the general requirements for wiring methods and materials; and by its scope, it can be modified by other articles within Chapter 3. Remember that per 90.3 the general requirements of chapters one through three can be modified by other articles in chapters five, six or seven. Only those sections in Article 300 that are referenced in Articles 725, 760, 770, 800, 805, 810, 820, 830, and 840, apply in the latter articles. The sections of Article 300 that apply to those later articles can be found under "Other Articles" which is typically in ".3" of the respective article. This is due to the unique nature of the requirements in the latter articles

Table Notes

Tables utilize a smaller size type for the footnotes, which are mandatory to the application of the table.

Exceptions

Basic rules are stated in standard (Roman) type and are followed by all the exceptions to the basic rule. Exceptions are set in italics. It is important to clearly read and understand the requirements of the basic rule before applying the exception. Exceptions apply only to the section or subsection they follow, unless stated differently in the rule. See the following example.

Several sections use a type of exception that can be considered "except as provided in _____ through_____." Look at 240.21 for an example of this concept. The general rule for location of the overcurrent device applies "except as specified in 240.21(A) through (H)."

Installation and Use

Section 110.3(B) states that equipment that has been certified by a Nationally Recognized testing laboratory (NRTL) to a product standard must be installed in accordance with any installation instructions that are included with its listing. [See the Informational Note to 110.3(C).] In some cases, the installation instructions can be more or less stringent than the *NEC* requirements.

PURPOSE

The *Code* is not intended as a design specification, and proper application dictates the use of nationally recognized product safety standards. Neither is the *Code* an instruction manual for untrained persons.

DEFINITION

Article 100, Definitions, generally contains definitions of terms only where used in two or more articles as stated in the Scope. When an article requires a definition of a term not used in other parts of the *Code*, that definition will be found within the article in which it is used at ".2" in the parallel numbering system. In most other cases, *Webster's Dictionary* or the *Institute of Electrical and Electronics Engineers Dictionary* will suffice. The cultivation of an electrical technical vocabulary is a must to understand the more intricate requirements. The *IEEE Standard Dictionary of Electrical Terms* proves helpful even to the seasoned veteran.

Applying the subtle rules that are used in writing the *Code* will assist old hands as well as the beginner. Failure to follow the simple rules leads to controversy and non-uniformity.

CHAPTER 3
CALCULATIONS

1 A 2-horsepower, 230-volt, single-phase, alternating-current motor is located 105 m (350 ft) from the service. What is the minimum size uncoated stranded copper conductor that is required to limit the voltage drop to 3 percent or less?

A. 12 AWG
B. 10 AWG
C. 8 AWG
D. 6 AWG

ANSWER _______________

PROCEDURE TO ESTABLISH ANSWER

The question is about voltage drop for a motor circuit.

1. In Index, find "Motors" under which find "Current, full load. see Full-load current motors."

2. Under "F" find "Full-load current motors," find "Alternating current" under which find "Single-phase, Table 430.248."

3. In Table 430.248 find a 2-horsepower, 230-volt motor rated 12 amperes.

4. Using the voltage drop formula:
   VD = 2 KLI/cm
   VD = 2 x K factor x Length of run x I
   (amperage) cm (circular mils of wire)

   In the formula, the "2" means out to the load and back from the load for the length of conductor. The "K" factor is the thermal conductivity multiplier with copper being 12.9 and aluminum at 21.2 (approximate value) at 75°C (167°F). The resistance of the conductors in Table 8 in Chapter 9 in "ohms per 1000 feet."

   WHERE:
   VD = Volts Dropped. You then take the volts dropped and divide it by the voltage of the circuit or system, and the result is the percentage (%) of voltage drop.

   A 3% voltage drop, for example
   = 3% x 230 volts
   = 6.9 volts

   K = 12.9
   cm = Circular mil area of conductor (Chapter 9, Table 8)
   VD = 2 x K factor x Length x amperes / cm area

When transposed, the new formula shown below will give us the minimum size wire needed to permit a maximum of 3% voltage drop on the circuit.

cm = 2 x K x L x I / VD
cm = 2 x 12.9 x 350 x 12/6.9
cm = 108,360 /6.9
cm = 15,704 cm

4. In Index find "Conductors," under which find "Copper, Properties, Chap. 9, Table 8."

5. In Table 8, we find 10 AWG has 10,380 circular mils and 8 AWG has 16,510 circular mils. Therefore, based on the above formula, we select an 8 AWG copper conductor.

6. VD = 2 x K x L x I / cm
   VD = 2 x 12.9 x 350 x 12
   VD = 108,360/16,510
   VD = 6.56 volts

7. VD = 6.56 volts ÷ 230 volts
   VD =.028 (or 2.8% voltage drop)
   The 8 AWG copper conductor will limit the voltage drop to 3% or less.

   The correct answer is C.

   For practice, better understanding and mastery of voltage drop calculations, the following exercises are suggested.

   Another method of finding the voltage drop is:

   Use steps 1 through 3 above, then:

8. In Index, find "Conductors," under which find "Copper, Properties, Chap. 9, Table 8."

9. In Table 8, we find the resistance of a 10 AWG uncoated stranded copper conductor to be 1.24 ohms per 1000 ft. The length of conductor to the load and back (2 x 350 feet) = 700 ft.

10. 700 ft./1000 ft = 0.7
    0.7 x 1.24 ohms = 0.868 ohms
    12. Use the Ohm's Law Formula
    VD (Volts Drop)
    E = I x R.
    E = 12 amps x 0.868 ohms
    E = 10.416 voltage drop

    Note: This is more than the allowed 6.9 volts (3 percent of 230 volts).

11. Per Table 8, the resistance of an 8 AWG copper uncoated stranded copper conductor is 0.778 ohms per 1000 ft.
    700 ft / 1000 ft = 0.7

0.7 x 0.778 ohms = 0.5446 ohms

12. Again, use the Ohm's Law Formula
    VD (Volts Drop)
    E = I x R.
    E = 12 amps x 0.5446 ohms
    E = 6.5352 volts

    Note: This is within the permitted 3 percent (6.9 volts drop) permitted in the question.

13. The correct answer is C.

2 A 208Y/120-volt, 3-phase, 4-wire wye-connected circuit to a process machine is 100 feet to the machine from the circuit breaker and has 45 amperes per phase of balanced resistance load. What is approximate percentage of voltage drop?

Note: Assume an ambient temperature of 75°C (167°F) and a conductor K (constant) factor of 12.9.

A. 1.5 percent
B. 1.8 percent
C. 2.6 percent
D. 3.0 percent

ANSWER ______________

PROCEDURE TO ESTABLISH ANSWER

The question is about calculating voltage drop.

K = 12.9
L = Length of circuit in feet (one way)
VD = Volt Drop
I = Current
Cm = Circular mil area of 6 AWG (Chapter 9, Table 8) = 26,240 cm
(6 AWG would be the required conductor size for the 45-ampere load.)

By using voltage drop formulas, use either of the following methods:

Note: it is very important to know how and where we get the formulas used, and how to use them. In the past, some engineers have used 2 and.866 in formulas for 3-phase voltage drop formulas. Many have wondered, Where did the .866 come from? The answer is: 2 x.866 = 1.732. You can choose to use 2 x.866 or just use 1.732 for 3-phase voltage drop calculations. For these calculations, we will use 1.732.

Method 1.

VD = 1.732 x KLI /cm
= 1.732 x 12.9 x 100 ft. x 45 amps
= 100,542.6 /26,240 cm (circular mil area of 6 AWG)
= 3.83 volts

The percentage of voltage drop is:

= 3.83 volts ÷ 208 volts

= 1.8%

The correct answer is B.

Method 2. By using the resistance in *NEC* Chapter 9, Table 8: (Note: The resistance of 6 AWG uncoated copper wire is.491 ohms per 1,000 feet.)

VD = IR x 1.73

Where R = 0.491/1000 x 100 (feet)

= 0.0491 ohms

VD = 45 amps x 0.0491 ohms x 1.73 for three phase

= 3.8 volts drop

Percent volt drop

= 3.8 Vd ÷ 208 volts

= 1.8% drop

Method 3. This method is not often used.

VD = 2 x R x L x I/1000

= 2 x 0.491 ohms x 100 ft. x 45 amp/1000 ft

= 4.419 volts

Percent volt drop

= 4.419 volts x 8.66

= 38.26854 volts

38.26854 volts/208 volts

= 0.1839833

=.0184

= 1.8%

3 What is the calculated load for three pieces of commercial electric cooking equipment that may be operated at the same time? The loads are 5 kVA, 3.5 kVA and 9 kVA, all operating at 240 volts single-phase.

A. 45 amperes
B. 54 amperes
C. 60 amperes
D. 66 amperes

ANSWER ______________

PROCEDURE TO ESTABLISH ANSWER

The question is about calculating branch-circuit loads for commercial cooking equipment.

1. In Index, find "Loads" under which find "Branch circuits" under which find "Calculations, Art 220."

2. Scan Article 220 and find Section 220.56 and Table 220.56, which for three units permits a 90 percent demand factor

3. Proceed with the calculation.

   = 5,000 VA + 3,500 VA + 9,000 VA
   = 17,500 VA

   = 17,500 VA x 90%
   = 15,750 VA

   = 15,750 VA ÷ 240 volts
   = 65.63 amperes (round up to 66 amps, per 220.5(B), Fractions of an Ampere).

4. The correct answer is D.

4 What size Type THW copper conductors installed in rigid metal conduit is required for a load consisting of 64,750 VA at 120/240 volts, single-phase, 3-wire?

A. 250 kcmil
B. 300 kcmil
C. 400 kcmil
D. 500 kcmil

ANSWER ______________

PROCEDURE TO ESTABLISH ANSWER

The question is about sizing of conductors.

1. Using the Ohm's Law formula

   I = P/E
   = 64,750 VA (Watts power) ÷ 240 V
   = 269.79 amperes.

2. In Index, find "Conductors" under which find "Ampacities, Tables 310.14 through 310.21, then go to 310.15 in Article 310 and find Table 310.16 , etc."

3. In Table 310.16 in the 75°C column (Type THW), find an ampacity that is equal to or next greater than 270 amperes, which is a 300-kcmil copper (rated at 285 amperes).

4. The correct answer is B.

5 For the purpose of determining conductor fill in conduit, what is the total area in square inches of three 4/0 AWG Type THWN and three 2/0 AWG Type THWN conductors?

A. 0.7823
B. 1.6380
C. 2.0329
D. 4.6298

ANSWER ______________

PROCEDURE TO ESTABLISH ANSWER

The question is about "area in square inches" of conductors.

1. In Index, find "Conduits," under which find "Conductors, number in, Chap. 9, Table 1, Annex C, Tables C1 through C-13(A)."

2. Note 6 of Table 1 states, "For combinations of conductors of different sizes, use the actual dimensions or Tables 5 and 5A [in Chapter 9] for dimension of conductors." Table 5A is used for compact copper and aluminum conductors. Because we have not any actual dimensions provided, we will use Table 5

3. In Table 5, find Type THWN where 4/0 AWG has an area of 0.3237 sq. in. and 2/0 AWG has an area of 0.2223 sq. in.

4. 3 conductors x 0.3237 sq. inches = 0.9711 sq. inches
   3 conductors x 0.2223 sq. inches = 0.6669 sq. inches

   Total Area = 0.9711 sq. inches + 0.6669 sq. inches
   Total Area = 1.6380 square inches

5. The correct answer is B

6

What is the amperage of a 975-watt load operating at a 75 percent power factor on a 120-volt ac single-phase circuit?

A. 4.2
B. 8.7
C. 10.8
D. 12.1

ANSWER ______________

PROCEDURE TO ESTABLISH ANSWER

The question is about calculating current.

1. Formulas to use are:
   a. For single-phase, I = P/E x PF, or I = W ÷ (E x PF)

2. Note: PF = Power Factor
   a. For 3-phase, I = W ÷ (E x 1.732 x PF)

3. Since the question involves single-phase load, go to single-phase formula.

4. I = 975 watts ÷ (120 volts x.75 PF)
   = 975 watts ÷ 90-volts
   = 10.8 amperes

5. The correct answer is C.

7 A 3-phase motor control center is fed with two sets of 500 kcmil Type THW copper conductors in parallel in a single raceway. What is the maximum allowable conductor ampacity if the conductors are installed in one metric designator 103 (4 in.) conduit?

A. 380 amperes
B. 608 amperes
C. 760 amperes
D. 800 amperes

ANSWER ______________

PROCEDURE TO ESTABLISH ANSWER

The question is about the ampacity of conductors where paralleled in a single raceway.

1. In Index, find "Ampacities" under which find "Conductors, 310.14 through 310.21 then go to Article 310 and find Tables 310.16 through 310.21."

2. Section 310.15(A) requires that ampacities for conductors rated 0–2000 volts shall be as specified in Tables 310.16 through 310.21.

3. In Table 310.16 find that 500-kcmil Type THW copper conductors have an ampacity rating of 380 amperes in the 75°C column.

4. Where six paralleled 500 kcmil conductors are contained in one conduit, Section 310.15(C)(1) applies and requires that the ampacity ratings given in Table 310.16 through 310.19 be reduced to 80 percent per Table 310.15(C)(1).

5. 380 amperes x (2 conductors per phase)
   I = 760 amperes x .8
   I = 608 amperes per phase.
   Note: 80% of the ampacity must be used because of 6 current-carrying conductors installed in the raceway together.

6. The correct answer is B.

8 An 800-ampere service switch is supplied from a wye-connected 480Y/277-volt utility transformer with the neutral point grounded. Service-entrance conductors are two sets of parallel 500 kcmil Type THWN conductors installed in the same raceway. There are no line-to-neutral loads. The minimum size copper grounded conductor required to be brought from the transformer to the service is:

A. 6 AWG
B. 4 AWG
C. 2 AWG
D. 2/0 AWG

ANSWER ______________

PROCEDURE TO ESTABLISH ANSWER

The question is about the minimum size permitted for the grounded conductor (neutral).

1. In Index, find "Conductors" under which find "See Grounded Conductors" under which find "Alternating current systems, 250.24 through 250.26." You could also find "Neutral' in Index and you will find "see also Conductors, grounded."

2. Section 250.24(C) states that the neutral (grounded) conductor of an ac supplied service shall be connected to each service disconnecting means. Section 250.24(C)(1) requires that the grounded conductor not be smaller than specified in Table 250.102(C)(1).

3. Section 250.102(C)(1) requires the grounded conductor to not be smaller than specified in Table 250.102(C)(1). The first two columns of Table 250.102(C)(1) provide the sizes for copper, aluminum and copper clad aluminum ungrounded conductors. The title of those two columns indicates the sizes are for the largest ungrounded conductor or the equivalent area for parallel conductors (AWG/kcmil). In our case, we are dealing with two 500 kcmil conductors per phase or an equivalent area of 1000 kcmil.

4. Table 250.102(C)(1) requires a 2/0 AWG copper grounded conductor for the 1000 kcmil (parallel 500 kcmil) phase conductors.

5. The correct answer is D.

9 The total inside area of a school is 160,000 square feet, and an outdoor-lighted play area is 120,000 square feet. The total connected load (including the air-conditioning), before applying demand factors is 1,740,000 VA. What is the total VA load for the school, after applying demand factors, as permitted using the optional method of calculation?

A. 1,740,000 VA
B. 1,425,000 VA
C. 480,000 VA
D. 1,305,000 VA

ANSWER ______________

PROCEDURE TO ESTABLISH ANSWER

The question is about optional method of calculations for services for a school.

1. In Index, find "Service-entrance conductors" under which find "Size, 230.42."

2. Section 230.42(A) requires service-entrance conductors to be of sufficient size to carry the load as calculated in Article 220.

3. Scan Article 220 for the optional method of calculation and find Section 220.86, "Optional Method — Schools."

4. The total connected load is 1,740,000 VA. The building square foot area is 160,000 square feet. The playground area is not considered since it is not required in the lighting calculations of Section 220.12 (floor areas by outside wall dimensions).

5. Volt-amperes per sq. ft.
   = 1,740,000 ÷ 160,000 sq. ft.
   = 10.875 VA per sq. ft.

In Table 220.86, apply demand factors as follows:

10.875 VA - 3 VA	= 7.875 VA
160,000 sq. ft x 3 VA x 1.00 (at 100%)	= 480,000 VA
7.875 VA x 160,000 sq. ft x .75 (at 75%)	= 945,000 VA
480,000 VA + 945,000 VA	= 1,425,000 VA

6. The correct answer is B.

10

A rigid metal conduit contains four 6 AWG Type THW, four 2 AWG Type THW, and eight 8 AWG Type THW copper conductors. What is the minimum permitted trade size for a single conduit containing these conductors?

A. Metric Designator 41 (1½ in.)
B. Metric Designator 53 (2 in.)
C. Metric Designator 78 (3 in.)
D. Metric Designator 91 (3½ in.)

ANSWER ______________

PROCEDURE TO ESTABLISH ANSWER

The question is about conductor fill in rigid metal conduit.

1. In Index, find "Conductor fill" under which find "Rigid metal conduit, 344.22."

2. Section 344.22 refers to Table 1, Chapter 9, under which Note 6 for combinations of conductors refers to the actual dimensions or Tables 5 and 5A for dimensions of conductors and Table 4 for applicable size of conduit. Since the question does not provide any actual dimensions, we will refer to the Tables.

3. Refer to Table 5 for conductor area,

6 AWG (THW) is 0.0726 sq. inches x 4 conductors	= .2904
2 AWG (THW) is 0.1333 sq. inches x 4 conductors	= .5332
8 AWG (THW) is 0.0437 sq. inches x 8 conductors	= .3496
Total of 16 conductors =	1.1732 sq. in.

4. Refer to Table 4 "Rigid Metal Conduit," under "Over 2 Wires 40%" column,

Metric Designator 41 (1½ in.) conduit	= 0.829 sq. in.
Metric Designator 53 (2 in.) conduit	= 1.363 sq. in.

5. A metric designator 53 (2 in.) conduit is required.

6. The correct answer is B.

11

Two 500 kcmil Type THW and six 4/0 AWG Type THHW conductors are installed in a sheet metal auxiliary gutter. What is the minimum size gutter permitted where the conductors are deflected?

A. 5" x 5"
B. 6" x 6"
C. 8" x 8"
D. 10" x 10"

ANSWER ______________

PROCEDURE TO ESTABLISH ANSWER

The question is about size of an auxiliary gutter containing deflected conductors.

1. In Index, find "Auxiliary gutters" under which find "Conductors" under which find "Number of, 366.22."

2. Section 366.22(A) states that the sum of the cross-sectional areas of all contained conductors at any cross section of a sheet metal auxiliary gutter shall not exceed 20 percent of the interior cross-sectional area of the sheet metal auxiliary gutter.

3. To determine that the 20 percent fill is not exceeded, it is necessary to determine the area of the conductors.

4. In Index, find "Conductors" under which find "Sizes, 110.6, Chap. 9, Tables 5 and 5A..." Table 5 is applicable since it provides conductor area in square inches for the type insulation addressed by the question.

5. Find the conductors and determine the area as follows:

   500 kcmil Type THW = 0.7901 sq. in. x 2 conductors
   = 1.5802 sq. in.

   4/0 AWG Type THHW = 0.3718 sq. in. x 6 conductors
   = 2.2308 sq. in.

   Total Area = 1.5802 sq. in. + 2.2308 sq. in.
   Total Area = 3.8110 sq. in.

6. Since the total area cannot exceed 20 percent or 1/5 of the gutter area, the minimum gutter area is 3.8110 sq. in. area of conductors x 5 inch= 19.055 square inches. (20% x 5 = 100%).

7. The 5-inch x 5-inch = 25 square inch gutter would be large enough for the eight conductors, but the deflection of the conductors must also be considered. (25 sq. in. is larger than 19.055 square inches).

8. In Index find "Deflection of conductors," under which find "Auxiliary gutters, 366.58(A)."

9. Section 366.58(A) requires that dimensions corresponding to 312.6(A) shall apply.

10. Section 312.6(A) requires that conductors not be deflected within a cabinet or cutout box, unless a gutter having a width in accordance with Table 312.6(A), is provided.

11. In the question, the largest conductor is 500 kcmil, which requires a minimum gutter width of 152 mm (6 in.) per Table 312.6(A). Therefore, a minimum 6" x 6" auxiliary gutter is required.

12. The correct answer is B.

12

Calculate the minimum size Type THW copper conductors required to serve the following motor loads from a 480Y/277-volt, 3-phase, 4-wire service: two 40-horsepower, 460-volt, 3-phase synchronous type, 90 percent PF and one 10-horsepower squirrel-cage, 460-volt, 3-phase motor at 100 percent PF.

A. 3 AWG
B. 6 AWG
C. 1 AWG
D. 1/0 AWG

ANSWER ______________

PROCEDURE TO ESTABLISH ANSWER

The question is about the minimum size conductors required for several motors.

1. In Index, find "Motors" under which find "Conductors" under which find "Ampacities, 430.6."

2. Section 430.6, Ampacity and Motor Rating Determination. Section 430.6(A)(1) states that the (FLC) values given in Tables 430.247 through 430.250, including notes, are to be used, instead of the actual current rating (FLA) marked on the motor nameplate to determine the ampacity of conductors, the ampere ratings of switches or the branch-circuit short-circuit and ground-fault protection.

3. Table 430.250 is applicable for 3-phase motors. The note below the table requires that for a 90 percent power factor (for synchronous motor rating), the table ratings be multiplied by 1.1.

4. Section 430.24 requires that conductors supplying two or more motors have an ampacity equal to the sum of the full load current (FLC) rating of all the motors plus 25 percent of the highest rated motor in the group.

5. Find in Table 430.250:

40-horsepower sync at 52 amps x 1.1 (for 90% PF)	= 57.2 amps x 2 motors	= 114.4 amps
52 amps x 25% (largest motor)		= 13 amps
10-horsepower squirrel cage		= 14.0 amps

Total = 114.4 amps (two 40 HP motors) + 13 amps (25% of largest motor - one of 40 HP] + 14.0 amps (10 HP motor)
= 141.4 amperes

6. In Index, find "Conductors" under which find "Copper" under which find "Ampacities, 310.14 through 310.21 and then go to Article 310 to access Table 310.16, etc."

7. Read down Type THW column in Table 310.16 under copper and find an ampacity rating equal to or greater than 141.4 amperes. Next higher ampacity listed is 150 amperes which is 1/0 AWG THW.

8. The correct answer is D.

Note: Section 220.5(B) under "Fractions of an Ampere," indicates calculations shall be permitted to be rounded to the nearest whole ampere, with decimal fractions smaller than 0.5 amperes can be dropped.

13

Which one of the following statements about Type THW conductors installed in electrical metallic tubing is NOT true?

A. Electrical metallic tubing containing three 1 AWG Type THW conductors shall be of a minimum trade size of metric designator 35 (1¼ in.)
B. The area of 1 AWG Type THW conductor is 0.1901 square inches.
C. The area of 8 AWG Type THW conductors is 0.0437 square inches.
D. A 8 AWG THW conductor can also be pulled into the metric designator 35 (1¼ in.) tubing with the three 1 AWG Type THW conductors.

ANSWER _____________

PROCEDURE TO ESTABLISH ANSWER

The question is about conductor fill in electrical metallic tubing.

1. In Index, find "Conductor fill" under which find "Electrical metallic tubing, 358.22."

2. Section 358.22 refers to Table 1, Chapter 9, under which Note 1 refers to Annex C for maximum number of conductors all of the same size; and Note 6 for combinations of conductors refers to the actual dimensions or Tables 5 and 5A for dimensions of conductors and Table 4 for applicable size of conduit.

3. Scan Annex C and find Table C.1 which permits three 1 AWG Type THW conductors in a metric designator 35 (1¼ in.) EMT conduit. Answer A is true.

4. In Table 5, find Type THW conductors and the respective areas for the conductors stated in the question. Answers B and C are true. Proceed to determine if the 8 AWG conductor can be installed in the metric designator 35 (1¼ in.) tubing.

   1 AWG is 0.1901 sq. in. x 3 conductors = 0.5703 sq. in.
   8 AWG is 0.0437 sq. in. x 1 conductor = 0.0437 sq. in.
   Total area = 0.6140 sq. in.

5. Referring to Table 4 "Electrical Metallic Tubing" under "Over 2 Wires 40%" column.

   Metric Designator 35 (1¼ in.) tubing = 0.598 sq. in. The total area of all the conductors is 0.614. Therefore, the 8 AWG conductor is not permitted to be pulled into the tubing with the 1 AWG conductors. Answer D is not true.

6. The correct answer is D.

14 Without considering any exceptions, what is the calculated volt-ampere number required to be used in sizing the conductor and overcurrent protection device for a feeder to a general lighting panel for a store building that has an overall dimension of 60' x 80' of which 400 square feet is used for storage?

A. 14,400 VA
B. 16,500 VA
C. 9,880 VA
D. 18,000 VA

ANSWER _______________

PROCEDURE TO ESTABLISH ANSWER

The question is about calculation used to size a feeder conductor and overcurrent protection device for a store building.

1. In Index, find "Lighting" under which find "Branch circuits, calculation of load, 220.12."

2. Go to Section 220.12 and then to Table 220.12.

3. Table 220.12 "superscript Note h" states a store is a retail occupancy. Table 220.12 shows a retail occupancy requires a minimum unit load of 1.9 VA for stores and storage spaces are not shown so would be added to the retail space calculation.

4. Scan Article 220 and find Table 220.12. It is assumed that the store lighting load would be considered to be continuous (on for 3 hours or more) and the storage load noncontinuous. In Index, find "continuous load," under which find "Applications, 210.19(A)(1), 210.20(A), 215.2(A)(1). Since this is a feeder, 215.2(A)(1) applies and requires the conductor size and rating of the overcurrent protection device for a feeder to be based on the noncontinuous load plus 125 percent of the continuous load. A note at the bottom of Table 220.12 states that "the 125 percent multiplier for a continuous load as specified in 210.20(A) is included when using the unit loads in this table for calculating the minimum lighting load for a specified occupancy." This means the 125% value is already figured into the table.

5. Proceed with the calculations:
   60' x 80' = 4800 sq ft total area
   4800 sq ft + 400 sq ft (storage) = 5200 sq ft of store
   5,200 sq. ft. at 1.9 VA per sq. ft. = 9,880 VA

   Lighting load total volt-amperes = 9,880 VA

6. The correct answer is C.

15

For branch-circuit or feeder calculations in industrial occupancies, what is the unit load for each outlet supplying heavy-duty lampholders that are not used for general illumination?

A. 180 volt-amperes
B. 600 volt-amperes
C. 20 amperes
D. 5 amperes

ANSWER _______________

PROCEDURE TO ESTABLISH ANSWER

The question is about heavy-duty lampholder unit loads.

1. In Index, find "Lampholders" under which find "Heavy duty. see Heavy-duty lampholders."
2. Under "Heavy-duty lampholders," find "Unit loads, 220.14(E)."
3. Section 220.14(E) requires 600 volt-amperes to be used for an outlet for a heavy-duty lampholder.
4. The correct answer is B.

16

Without considering any exceptions, which one of the following statements about calculated general lighting loads is true?

A. The load for a 10,000 square foot office building is 40,000 volt-amperes.
B. The load for a 10,000 square foot school is 40,000 volt-amperes.
C. The load for a 10,000 square foot restaurant is 10,000 volt-amperes.
D. The load for a 10,000 square foot church (Religious facility) is 22,000 volt-amperes.

ANSWER _______________

PROCEDURE TO ESTABLISH ANSWER

The question is about calculating lighting loads for various buildings.

1. In Index, find "Lighting" under which find "Branch circuits, calculation of load, 220.12."
2. Section 220.12 refers to Table 220.12. The lighting demand for the buildings listed is:

Offices	= 1.3 VA per sq. ft.
Schools	= 3 VA per sq. ft.
Restaurants	= 1.5 VA per sq. ft.
Churches (Religious facility)	= 2.2 VA per sq. ft.

3. Calculate the lighting load for each building:

Office – 10,000 sq. ft. at 1.3 VA	= 13,000 VA
School – 10,000 sq. ft. at 3 VA	= 30,000 VA
Restaurant – 10,000 sq. ft. at 1.5 VA	= 15,000 VA
Church (Religious facility) – 10,000 sq. ft. at 2.2 VA	= 22,000 VA

4. The correct answer is D.

   Note: The lighting load in Table 220.12 is considered to be continuous. The 125 percent multiplier for a continuous load as specified in 210.20(A) is included when using the unit loads in this table for calculating the minimum lighting load for a specified occupancy.

17 A 1,500 square foot jewelry store containing the items listed below is supplied by a 120/240-volt service.

6.0 m (20 ft) of show window
Twenty-one receptacles with continuous duty operation
Ten 150-volt-ampere lamps in heavy-duty lampholders
Ten fluorescent luminaires with two 32-watt lamps each (Ballast amps =.69 amps /each)
One 1800 volt-ampere sign

What is the minimum number of 20-ampere-branch circuits required for this store?

A. 12 circuits
B. 6 circuits
C. 8 circuits
D. 10 circuits

ANSWER ______________

PROCEDURE TO ESTABLISH ANSWER

The question is about the minimum number of branch circuits required for a store. Since these calculations will involve several *Code* sections and detailing each step and reference would be lengthy, only the procedure is shown.

1. Section 210.11(A) requires the minimum number of branch circuits to be determined from the total calculated load and size of circuits used. Also, the load on any circuit is not to exceed the maximum specified by Section 220.18. For continuous load Section 210.19(A)(1)(a) requires the circuit size to be not less than 125 percent of the continuous load. Refer to Section 220.14 for other minimum loads.

2. Show window outlets: 220.14(G)(2) = 20 feet x 200 VA per linear foot x 125%
   = 5000 VA ÷ 120V
   = 41.67 amps

   Note that a continuous load on any branch circuit cannot exceed 80% of the branch-circuit rating of a circuit breaker, unless the breaker is rated for 100% duty. (20 amps x .80 = 16 amps).
   = 41.67 amps / 16 amps
   = 2.60 circuits (three 20-amp circuits)

   Note: Section 220.14(G)(1) permits "the unit load per outlet as required in other provisions of this section." This amounts to 180 VA per receptacle outlet for illumination or other purposes and ties in with Section 210.62, Show windows. At least one receptacle outlet is required to be installed for every 12 linear feet of show window, maximum, or fraction thereof. This provision in the Code is used often for show windows in a pet store, for example, where a heat lamp or small fan may be used to keep puppies comfortable, or a small motor is used to keep a fish tank aerated.

Twenty-one receptacles: 220.14(I)
21 receptacles x 180 VA each = 3,780 VA.
Receptacles must be figured at 125 percent for continuous duty since the information given in the question states that this is the case; therefore, calculate

= 3780 VA receptacle load x 1.25 for continuous use
= 4725 VA ÷ 120V
= 39.38 amps / 16 amps (Note: 20 amps x .80 = 16 amps).
= 2.46 circuits (three 20-amp circuits).

Ten 150-watt heavy duty lamps: 220.14(E)

=10 lamps x 600 VA each
= 6000 VA X 1.25 ÷ 120V
= 62.5 amps / 16 amps
= 3.9 circuits (four 20-amp circuits).

Ten fluorescent luminaires: 220.18(B)

= 10 luminaires x .69 amps per ballast x 1.25
= 8.6 amps (one 20-amp circuit)

One 1800 volt-amperes sign: 220.14(F); 600.5

= 1800 VA ÷ 120V
= 15 amps (one 20-amp circuit)

3. Total number of 20-ampere branch circuits required:

Show window outlets	= Three 20-ampere circuit
Twenty-one receptacles	= Three 20-ampere circuits
Ten 150-watt lamps	= Four 20-ampere circuit
Ten fluorescent luminaires	= One 20-ampere circuit
One 1800 volt-amperes sign	= One 20-ampere circuit
Total	= Twelve 20-ampere circuits

4. The correct answer is A.

18 How many 300-watt industrial process lamps may be installed on a branch circuit rated at 15 amperes, 120 volts?

A. 3
B. 4
C. 5
D. 6

ANSWER ______________

PROCEDURE TO ESTABLISH ANSWER

The question is about maximum load on a branch circuit.

1. In Index, find "Branch circuits" under which find "Maximum loads, 220.18" and under "Branch Circuits" find "Ratings, 210–Part II."

2. Scan Article 210, Part II, and find Section 210.19(A)(1)(a), which requires that the rating of a branch circuit serving continuous loads, such as industrial loads, shall not be less than the noncontinuous load plus 125 percent of the continuous load.

3. Determine the volt-ampere capacity of the circuit by multiplying 120 V x 15 amps = 1800 VA.

4. The lamps must be calculated as a continuous load since these industrial process lamps are used in an industrial occupancy.

   300 watts each lamp (VA) x 1.25 (125%)
   = 375 VA
   = 1800 VA ÷ 375 VA
   = 4.8 or 4 lamps on a 15-ampere circuit.

5. The correct answer is B.

19 A feeder supplies a total connected continuous load of 4800 volt-amperes. The overcurrent protection device is not listed for 100 percent operation. What is the calculated volt-ampere load for the selection of the overcurrent protection device?

A. 3200
B. 4800
C. 5400
D. 6000

ANSWER ______________

PROCEDURE TO ESTABLISH ANSWER

The question is about calculation of continuous loads.

1. In the Index find "Feeders" under which find "Calculation of loads" "see loads."

2. In Index find "Loads" under which find "Feeders," under which find "Calculations for, 215.2(A), Article 220, Annex D."

3. Section 215.2(A)(1)(a) requires feeder conductors to have sufficient ampacity to supply the load served at 125% for continuous loads and 100% for noncontinuous loads for feeder conductors.

4. For feeder overcurrent protection, Section 215.3 states that where a feeder supplies a continuous load, the overcurrent device shall be not less than 125 percent of the continuous load.

5. For a 4800-volt-ampere continuous load:
   4800-volts x 1.25 = 6000 volt-amperes.

6. The correct answer is D.

20 An existing 480Y/277-volt, 3-phase/4-wire feeder to a load of electric-discharge lighting was supplied from a 125-ampere overcurrent device. The phase conductors and neutral were 1 AWG Type THW copper with a 6 AWG equipment grounding conductor. Due to voltage drop on the feeder circuit, the ungrounded conductors had to be increased to 2/0 AWG copper. The minimum size grounded conductor and equipment grounding conductors are:

A. 1 AWG and 6 AWG
B. 1 AWG and 4 AWG
C. 2/0 AWG and 6 AWG
D. 2/0 AWG and 4 AWG

ANSWER ______________

PROCEDURE TO ESTABLISH ANSWER

The question is about sizing neutral and equipment grounding conductors.

1. In Index find "Neutral," under which find "Feeder load, 220.61" which states that there is no reduction of neutral capacity for nonlinear loads, such as electric-discharge lighting. Based on Section 310.15(E)(3) and 220.61, for this type of load, the neutral is considered to be a current-carrying conductor and the neutral would have to be increased to 2/0 AWG, as was done for the phase conductors.

2. In Index find "Equipment grounding conductors," under which find "Sizing , 250.122, 250.190(C)(3)."

3. Section 250.122(B) provides that where conductors are adjusted in size to compensate for voltage drop, equipment grounding conductors, where required, shall be adjusted proportionately according to circular mil area increase of the ungrounded conductors.

4. Refer to Chapter 9, Table 8 for circular mil area of conductors.
   1 AWG Neutral size before increase = 83,690 cm area
   6 AWG Equipment grounding conductor size before increase = 26,240 cm area
   2/0 AWG Neutral size after increase = 133,100 cm area

5. Determine the ratio of the ungrounded and grounded conductors.
   133,100 cm (2/0 AWG)/83,690 cm (1 AWG) from Table 8 Chapter 9
   = 1.59 ratio

6. Multiply area of required equipment grounding conductor by the determined ratio:
   26,240 cm of 6 AWG x 1.59 ratio = 41,721.6 minimum cm area.

7. Per Table 8 in Chapter 9, circular mils of a 4 AWG conductor are 41,740, which would be large enough for the required increased ratio (41,721 cm).

8. The correct answer is D.

21

A restaurant served by a 208Y/120-volt, 3-phase, 4-wire service has the following kitchen equipment (all appliance loads are 3-phase):

One 2.5 kVA dishwasher
One 2 kVA dishwasher booster heater
One 6 kVA water heater
Two 12 kVA ranges
Two 5 kVA deep-fryers
Two 2 kVA toasters
One 2 kVA warming oven

The restaurant also has lighting loads of 4 kVA (includes 125 percent for continuous load) and sixty-seven general purpose receptacles totaling 12 kVA. What is the minimum size of Type THW copper feeder conductor required to supply the kitchen panelboard?

A. 2 AWG
B. 1 AWG
C. 1/0 AWG
D. 2/0 AWG

ANSWER ______________

PROCEDURE TO ESTABLISH ANSWER

The question is about calculating the minimum size of kitchen feeder conductors.

1. In Index, find "Kitchen equipment, commercial, 220.56, 422.11(F)(2)."

2. Section 220.56 permits calculating the load for commercial electric cooking equipment, dishwasher booster heaters, water heaters and other kitchen equipment according to Table 220.56.

3. Table 220.56 demand factors are to be applied to all equipment that has either thermostatic control or intermittent use as kitchen equipment, except heating, ventilating or air conditioning. Since ten items apply, a 65 percent demand factor can be applied.

Dishwasher	= 2.5 kVA
Booster heater	= 2.0 kVA
Water heater	= 6.0 kVA
Two ranges at 12 kVA	= 24.0 kVA
Two deep-fryers at 5 kVA	= 10.0 kVA
Two toasters at 2 kVA	= 4.0 kVA
Warming oven	= 2.0 kVA

Add all these loads, and the total without demand factor

= 50.5 kVA before demand factor

Kitchen equipment load:

= 50.5 kVA x 65% demand factor applied
= 32.8 kVA after demand factor

4. Per Table 220.44 for 12 kVA receptacle load:

First 10 kVA @ 100%	= 10.0 kVA
50% of remaining 2 kVA	= 1.0 kVA
Lighting load (already at 125%)	= 4.0 kVA
Plus the kitchen equipment load	= 32.8 kVA after demand factor
Total load	= 47.8 kVA

5. 47.8 kVA x 1000 = 47,800 VA
   I = VA/E x 1.73 (square root of 3)
   I = 47,800 VA/208 volts x 1.73
   I = 47,800 VA/360-volts
   I = 132.78 amperes

6. Referring to Table 310.16 for conductor ampacity, scroll down the Type THW column in the 75° column to the ampacity next greater than 132.78 amperes, which is 150 amperes. Read left to find a 1/0 AWG Type THW copper conductor is required to supply this load.

7. The correct answer is C.

22 When using the optional calculation method for calculating the service size for a 100-unit multifamily dwelling with each unit having an electric range and electric space heating, what percent demand factor is permitted to be applied to the electric space heating load?

A. 23%
B. 45%
C. 65%
D. 70%

ANSWER ______________

PROCEDURE TO ESTABLISH ANSWER

The question is about using the optional method of calculating the service size for a multifamily dwelling building. Note: Service size is based on feeder calculations.

1. In Index find "Dwellings," under which find "Feeder load, calculations for, 220.14(J), 220.16(A), 220.82 through 220.85, Annex D."

2. Scan these sections and find Section 220.84, that is applicable for calculating multi-family service or feeder loads where specific conditions are met. Demand factors in Table 220.84 are permitted to be used.

3. Section 220.84(C)(3) permits applying the demand factors to the nameplate rating of the ranges and 220.84(C)(5) allows a demand factor to be applied to the larger of the air-conditioning load or space heating load.

4. Table 220.84 lists a demand factor of 23 percent for dwellings with the number of units equal to or over 62. This is a 100-unit project.

5. The correct answer is A.

23 What is the largest standard ampere rating for cartridge fuses?

A. 600
B. 1000
C. 1200
D. 6000

ANSWER ______________

PROCEDURE TO ESTABLISH ANSWER

The question is about code-recognized standard fuse ratings.

1. In Index find "Fuses," under which find "Ratings, 240.6."

2. Section 240.6(A) lists the largest standard rating as 6000 amperes.

3. The correct answer is D.

24 The ampacity of a 2-wire, 18 AWG Type SO flexible cord is:

A. 7 amperes
B. 10 amperes
C. 15 amperes
D. 20 amperes

ANSWER ______________

PROCEDURE TO ESTABLISH ANSWER

The question is about the rated ampacity of flexible cord.

1. In Index find "Cords" and then find " Flexible, Article 400., 422.16, 422.43" Also find in the Index "Tables," under which "Ampacities," under which find "Flexible cord, Table 400.5(A)(1) through (A)(3)."

2. Table 400.5(A)(1) under the third column lists Type "SO," and provides a superscript letter [a] by subheading A and superscript letter [b] subheading B.

3. The notes following the table indicate that subheading A applies to cords with three current-carrying conductors, and subheading B applies to cords with two current-carrying conductors, which is applicable to the question.

4. Read down the AWG column to 18 and across to column B for a rating of 10 amps.

5. The correct answer is B.

25 Which one of the following restaurants is NOT considered to be a place of assembly?

A. One with a capacity of 50 persons
B. One with a capacity of 100 persons
C. One with a capacity of 125 persons
D. One with a capacity of 150 persons

ANSWER ______________

PROCEDURE TO ESTABLISH ANSWER

The question is about classification of places of assembly.

1. In Index find " Assembly occupancies " under which find "Classifications, 518.2."
2. Sections 518.1 and 518.2 both confirm that an assembly occupancy is a building or portion of a building intended for the assembly of 100 or more persons.
3. The correct answer is A, not an assembly occupancy.

26 Which one of the following sets of paralleled Type THW conductors is permitted as the service-entrance conductors for a service with 2000-ampere main fuses?

A. Five sets of 500 kcmil copper conductors in five separate conduits
B. Five sets of 600 kcmil copper conductors in five separate conduits
C. Five sets of 600 kcmil aluminum conductors in five separate conduits
D. Five sets of 750 kcmil aluminum conductors in five separate conduits

ANSWER ______________

PROCEDURE TO ESTABLISH ANSWER

The question is about size of service-entrance conductors and where the next higher overcurrent device may be used.

1. In Index find "Service-entrance conductors," under which find "Overcurrent protection," under which find "230.90 through 230.92, 230.208."
2. Section 230.90(A) requires overload protection by an overcurrent device in series with each ungrounded service conductor having a rating not higher than the ampacity of the conductor.
3. Exception No. 2 recognizes fuses with a rating conforming to Section 240.4(B), which permits the next higher standard overcurrent device, etc., providing it does not exceed 800 amperes.
4. Determine the total ampacity of each of the above sets of conductors from Table 310.16.

500 kcmil Type THW copper	= 380 amps x 5 sets of conductors	= 1900 amps
600 kcmil Type THW copper	= 420 amps x 5 sets of conductors	= 2100 amps
600 kcmil Type THW aluminum	= 340 amps x 5 sets of conductors	= 1700 amps
750 kcmil Type THW aluminum	= 385 amps x 5 sets of conductors	= 1925 amps

5. Since the fuse rating (2000 amperes) cannot exceed the conductor ampacity, only the 600 kcmil copper conductors would be acceptable.

6. The correct answer is B.

   Note: Although Section 230.90(A) has other exceptions, the question does not provide any information to apply these other exceptions.

27

Which one of the following continuous rigid metal conduit raceway installations between pull points is in compliance?

A. A raceway with three 90-degree and two 45-degree bends
B. A raceway with three 90-degree and three 45-degree bends
C. A raceway with three 90-degree and four 30-degree bends
D. A raceway with two 90-degree, two 45-degree, and four 30-degree bends

ANSWER ______________

PROCEDURE TO ESTABLISH ANSWER

The question is about the maximum number of bends permitted in a conduit run.

1. The question is about the maximum number of bends permitted in a conduit run

2. In Index find "Rigid metal conduit," under which find "Bends, 344.24, 344.26, Chapter 9, Table 2."

3. Section 344.26 permits the equivalent of 360 degrees maximum between pull points.

4. The correct answer is A.

28

A 120/240-volt, single-phase, 3-wire service is installed for a multifamily dwelling. Using the standard calculation method, what is the minimum service size in amperes for a building having 40 dwelling units of 1000 square feet each with an 8 kVA range, 1.2 kVA dishwasher, 0.6 kVA disposer, 120-volt single-phase A/C at 18 amps and a 2.5 kVA, 240-volt water heater?

A. 1200 amperes
B. 1400 amperes
C. 1500 amperes
D. 1600 amperes

ANSWER ______________

PROCEDURE TO ESTABLISH ANSWER

The question is about calculating the service-entrance size for a multifamily dwelling.

1. In Index find "Service-entrance conductors," under which find "Size, 230.42."

2. Section 230.42(A) requires that the conductors be of sufficient size to carry the loads as calculated in accordance with Article 220.

3. Refer to Article 220 and find 220.14(J)(1), (J)(2), and (J)(3) .

4. Proceed with the calculations.

General lighting [220.14(J)(3):
= 1000 sq. ft. per unit x 3 VA per sq. ft.
= 3000 VA x 40 units
= 120,000 VA total per square foot

Kitchen small appliance [(210.11(C)(1) & 220.52]:
1500 VA per small appliance branch circuit x 2 circuits per unit x 40 units = 120,000 VA for small appliance branch circuits

Total general lighting and small appliance load
120,000 VA + 120,000 VA = 240,000 VA total lighting and small appliance loads

Section 220.52 permits Table 220.42 demand factor for this load.
First 3000 VA at 100% = 3,000 VA
(Note: 240,000 total - 3,000 VA at 100% = 237,000 VA)
Next 117,000 at 35% = 40,950 VA
(Note: 237,000 VA (Total minus 3000 VA)- 117,000 VA taken at 35% = 120,000 VA remainder)
Remaining 120,000 VA at 25% = 30,000 VA

Net General lighting and small appliance load
3,000 VA + 40,950 VA + 30,000 VA = 73,950 VA

Ranges (Table 220.55 – Column C) 40 Ranges
15 kVA + 1 kVA x 40 units from Table 220.55 demand factors = 55,000 VA

Appliances 220.53:

Dishwashers at 1.2 kVA x 40 units = 48,000 VA
Disposer at 0.6 kVA x 40 units = 24,000 VA
Water heater at 2.5 kVA x 40 units = 100,000 VA
Sub-Total for appliances = 48,000 VA+ 24,000 VA + 100,000 VA totals from above
= 172,000 VA Subtotal of appliances

Net Appliance Load (220.53)(doesn't apply to household cooking, clothes dryers, space heating and air conditioning

= 75% demand x 172,000 VA = 129,000 VA

40 air conditioners at = 18A x 120V = 2,160 VA per unit
= 2,160 VA x 40 units
= 86,400 VA subtotal of air conditioners

25% largest motor [220.18(A)], Assume A/C
= 2,160 x 25%
= 540 VA 25% for largest motor

Total calculated load: 73,950 VA (general lighting and small appliance loads) + 55,000 VA (ranges) + 129,000 (appliance loads) + 86,400 VA (A/C) + 540 VA (25%)
= 344,890 VA total load

5. Minimum service size is = 344,890 VA ÷ 240V
= 1,437 amperes.

6. The correct answer is C for next higher at 1500 amperes.

29

Using the diagram below, what is the approximate percent of volt drop using 12 AWG copper?

Note: For this problem, we will assume an ambient temperature of 75°C (167°F) and a conductor K (constant) factor of 12.9.

A. 1.98%
B. 2.48%
C. 3.0%
D. 5.0%

ANSWER ______________

PROCEDURE TO ESTABLISH ANSWER

The question is about calculating voltage drop.

1. The solution is as follows:
   K - 12.9
   L = length of circuit in ft. (one way)
   VD = volt drop
   I = current
   cm = circular mil area of conductor
   (Chapter 9, Table 8)

2. VD = 2 x K x L x I / cm
   2 x 12.9 x 100 feet x 12 amps/6530 cm for 12 AWG copper
   = 30,960/6530 cm
   = 4.74 volts dropped

3. Percent voltage drop
   = 4.74 volts ÷ 240 volts
   = 1.98%*

4. Check by using VD = IR
   I = 12 amps
   R = Resistance of 60 m (200 ft) of 12 AWG stranded copper conductor

R =.00198 ohms per foot (1.98 ohms per 1,000 feet ÷ 1,000
= 0.0198 ohms per foot)
(Chapter 9, Table 8)
=.00198 ohms per foot x 200 feet
= 0.396 ohms for 200 ft

VD = IR
= 12 amps x 0.396 ohms
= 4.75 volts drop

5. Percent volt drop
= 4.75 volts ÷ 240 volts
= 0.01979
= 1.98%

6. The correct answer is A.

Note: Section 210.19(A) Informational Note No. 3 recommends 3 percent maximum volt drop for branch circuits.

* 240 volts is used to determine the percent voltage drop, since , based on the diagram, the 12 amperes flow through the two-phase wires and both loads across 240 volts with no current in the neutral. If the type of loads and circuit switching were such that only one of the loads was energized at one time, the 12 amperes would then flow through one phase conductor and the neutral across 120 volts. In that case, the percent voltage drop should be based on 120 volts and would be 3.96 percent.

30

What is the general lighting load of an office building with an area of 4,500 square feet and the actual number of general-purpose receptacle outlets is unknown?

A. 13,500 VA
B. 15,750 VA
C. 18,000 VA
D. 10,350 VA

ANSWER ____________

PROCEDURE TO ESTABLISH ANSWER

The question is about the calculation of branch-circuit lighting loads.

1. In Index find "Branch circuits," under which find "Calculation, 220 Part II."

2. Scan Part 220 Part II, and find Section 220.12, that contains Table 220.12.

3. Proceed with the calculations.

Lighting load [Table 220.12]:
4500 sq. ft. x 1.3 VA per sq. ft. = 5850 VA

Receptacle outlets [see 220.14(K)(2)]:
4500 sq. ft. x 1 VA per sq. ft. = 4500 VA

General Lighting Load = 5850 VA + 4,500 VA
General Lighting Load = 10,350 VA

(the receptacle load may be added onto the lighting load at a value of 1 volt-ampere per square foot when the actual number of receptacles is not yet known.

4. The correct answer is D.

Note: The lighting load from Table 220.12 is already figured at 125% and the convenience receptacle load is noncontinuous.

31

Calculate the minimum size feeder overcurrent protective device for an office space of 50' x 100' which has a 120/240-volt, single-phase supply. It also contains 60 m (200 ft) of fixed multioutlet assembly where appliances are unlikely to be used simultaneously, two 16.7-amp, 120-volt copying machines, and one air-conditioner of 54.5 amps at 240 volts.

A. 175 amperes
B. 125 amperes
C. 225 amperes
D. 250 amperes

ANSWER ______________

PROCEDURE TO ESTABLISH ANSWER

The question is about minimum size of a feeder overcurrent device.

1. In Index find "Feeders," under which find "Calculation of loads, see Loads" Under Loads, find "Feeders, calculations for, 215.2(A), Article 220, Annex D."

2. Article 220 provides calculations, 220.40 references Part III deals with feeders and services. See 220.12 and Table 220.12 for lighting loads.

3. Calculate the lighting load based on Table 220.12.
   50 ft. x 100 ft. = 5000 sq. ft
   5000 sq. ft. x 1.3 VA per sq. ft. = 6,500 volt-amperes

   I = 6,500 volt-amperes ÷ 240 V
   I = 27.08 or 27 amperes (continuous lighting load)

4. Calculate the noncontinuous load.
   60 m (200 ft) multioutlet assembly in 220.14:
   180 VA per each 1.5 m (5 ft) of multioutlet assembly [220.14(H)(1)]
   = 200 ft ÷ 5 ft
   = 40 outlets x 180 VA
   = 7200 VA

= 7200 VA ÷ 120-V
= 60.0 amp at 120 V for multioutlet assembly

220.14(A)
Two copy machines at 16.7 amps each = 33.4 amp at 120 V

Total noncontinuous calculated 120 V load 60 amps for multioutlet Assembly plus copy machines at 33.4 A= 93.4 amp at 120 V or 46.7 amp at 240 V

Air-conditioning 54.5 amps x 125% = 68.1 amp at 240 V

Total noncontinuous calculated 120 V load
60.0 amperes of multioutlet assembly + 34.4 amperes for copy machines= 93.4 amp at 120 V (or 46.7 amp at 240 V)

5. Per 215.3, the minimum rating of a feeder overcurrent device is to be not less than the noncontinuous load plus 125 percent of the continuous load: (The only continuous load is the lighting with Table 220.12 already using 125%)

27 amps of lighting load

(Noncontinuous Load)
= 93.4 amps (60.0 amperes of multioutlet assembly + 34.4 amperes for copy machines)

Therefore, 27 amps + 93.4 amps
= 120.4 amps

6. With a 120.4 ampere calculated load, the minimum feeder overcurrent device rating is 125 amperes per Section 240.6, which provides the standard rating of overcurrent devices. Section 240.4(B) allows the use of the next standard size overcurrent device above the ampacity of the conductors where the circuit is less than 800-amperes and the ampacity of the conductors does not correspond with the standard ampere rating of fuse or circuit breaker.

7. The correct answer is B.

32 What is the minimum ampacity required for a 480Y/277-volt, 3-phase, 4-wire feeder supplying a 150 kVA fixed electric space heating load (assumed to be noncontinuous)?

A. 181 amperes
B. 226 amperes
C. 313 amperes
D. 391 amperes

ANSWER ______________

PROCEDURE TO ESTABLISH ANSWER

The question is about the minimum calculated feeder load for fixed electric space heating.

1. In Index find "Feeders," under which find "Calculation of loads, see Loads.." Under "Loads", find "Feeders, calculations for, 215.2(A), Article 220, Annex D."

2. Look in Part III of Article 220 and find that 220.40 references Part III of Article 220.. Section 220.40 requires feeder conductors to have sufficient ampacity to supply the load served and in no case shall the computed load of a feeder be less than the total computed branch-circuit load per Part II after demand factors of III or IV have been applied.

3. Scan Part III and find Section 220.51, that generally requires the feeder load to be calculated at 100 percent of the total connected fixed electric space-heating load.

4. Based on Step 3, the total feeder load is 150 kVA x 1000 = 150,000 VA at 100 percent.

5. The minimum conductor ampacity for the 3-phase load is:

   I (amps) = VA/E (volts) x 1.73
   I = 150,000 VA/480 volts x 1.73
   I = 150,000 VA /830.4
   I = 180.7 amperes
   I = 181 amperes, rounded up

6. The correct answer is A.

33

What is the minimum size Type THWN copper conductor required for a feeder supplying ten 12 kVA, 240-volt, single-phase electric ranges in a commercial restaurant kitchen?

A. 2 AWG
B. 1 AWG
C. 350 kcmil
D. 400 kcmil

ANSWER ______________

PROCEDURE TO ESTABLISH ANSWER

The question is about calculation of feeder loads for electric ranges in a commercial restaurant.

1. In Index find "Feeders," under which find "Kitchen equipment, commercial, 220.56."

2. Section 220.56 permits the load for commercial kitchen equipment to be in accordance with Table 220.56.

3. Per Table 220.56, a 65 percent demand factor can be applied to six or more units of equipment (electric ranges).

4. Proceed with the calculations:
   10 appliances at 12 kVA each = 120 kVA
   = 120 kVA x 65% demand factor = 78 kVA or 78,000 VA after demand factor

   = 78,000 VA ÷ 240 V
   = 325 amperes

5. Refer to Table 310.16 and find that a 400-kcmil Type THWN copper conductor is rated for 335 amperes, while a 350-kcmil Type THWN copper conductor is only rated for 310 amperes.

6. The correct answer is D.

34 What is the feeder neutral current for a multifamily dwelling supplied with a 120/240-volt, 3-wire service with a calculated load (including demand factors as per Table 220.42) of 111,450 VA, including a 40 kVA range load per Table 220.55?

A. 385 amperes
B. 432 amperes
C. 485 amperes
D. 500 amperes

ANSWER ______________

PROCEDURE TO ESTABLISH ANSWER

The question is about the load on a feeder neutral conductor.

1. In Index find "Neutral," under which find "Feeder load, 220.61."

2. Section 220.61 permits the unbalanced neutral load to be considered as 70 percent of calculated range load or a 70 percent demand permitted for that portion of the feeder load in excess of 200 amps.

3. Calculate the amperage of the respective loads.

   Total feeder load of 111,450 volt-amperes less range load of 40,000 = 71,450 VA. Proceed to convert the volt-ampere load to amperes:

   71,450 VA/240 volts without range = 298 amperes

   40,000 VA/240 volts range loads = 167 amperes

4. Total neutral load of feeder = general neutral load plus range load (including demand factor)
   298 amps without range + 167 amps
   = 465 amperes.

5. 100% of the first 200 A = 200 amperes

   70% of load in excess of 200 A: = 465 A – 200 A = 265 A x 70%
   = 185.5 Amps demand plus 200 amps at 100%
   =385.5 amps

6. The correct answer is A.

35 How many amperes will each phase conductor carry with a 240-volt, 3-phase, 15 kVA heater?

A. 36.1
B. 62.5
C. 72.3
D. 125

ANSWER ______________

PROCEDURE TO ESTABLISH ANSWER

The question is about calculating phase current on a 3-phase circuit.

1. Proceed with calculation:

   $P = E \times I$

   kVA = E (volts) x I (amps) x 1.73

   15,000 VA = 240 V x I x 1.73

   Therefore,

   I = kVA/V x 1.73

   I = 15,000 VA/240-V x 1.73

   I = 15,000 VA/415.2 $V_{effective}$

   I = 36.1 amperes

2. The correct answer is A.

36 What is the minimum size Type THW conductor required for a feeder supplying a balanced 100 kVA lighting actual load at 3-phase, 480Y/277 volt?

A. 1 AWG copper
B. 1/0 AWG copper
C. 2/0 AWG aluminum
D. 3/0 AWG aluminum

ANSWER ______________

PROCEDURE TO ESTABLISH ANSWER

The question is about determining conductor size for a specific load.

1. Calculate the load current.

   $I = P/E \times 1.73$

   I = E (volts) x 1.73

   I = 100,000 VA/480 volts x 1.73

   I = 100,000 VA/830.4 $V_{effective}$

   I = 120 amperes

2. It is assumed that an actual lighting load that is supplied from a 480Y/277-volt system would be considered to be a continuous load (on for 3 hours or more). Section 215.2(A)(1)(a) requires the conductor size for a feeder to be based on the noncontinuous load plus 125 percent of the continuous load.

3. 120 amperes x 125% = 150 amperes

4. Refer to Table 310.16 and find that a 1/0 AWG Type THW copper conductor in the 75° column is rated at 150 amperes. Although a 3/0 AWG aluminum conductor is rated at 155 amperes, the question asked for the minimum size conductor and, in accordance with 110.5, where the conductor material is not specified, the sizes given in the *NEC* are applied to copper conductors.

5. The correct answer is B.

37

A single-phase electric furnace rated at 10 kVA at 240 volts is connected to a single-phase, 208-volt system. What will be the output rating of the furnace?

A. 5.8 kVA
B. 6.9 kVA
C. 7.2 kVA
D. 7.5 kVA

ANSWER ______________

PROCEDURE TO ESTABLISH ANSWER

The question is about calculating the output of a 240-volt furnace operating at 208 volts.

1. Using the VA and voltage rating of the furnace determine the resistance of the furnace.

$R = E \times E / P$

$R = E \times E / kVA$

$R = 240\ \text{V} \times 240\ \text{V}/10{,}000\ \text{VA}$

$R = 57{,}600\ \text{V}/10{,}000\ \text{VA}$

$R = 5.76\ \text{ohms}$

2. Using the resistance of the furnace and voltage of the circuit determine the VA output.

$VA = E \times E/R$

$\text{VA} = 208\ \text{V} \times 208\ \text{V}/\ 5.76\ \text{ohms}$

$\text{VA} = 43{,}264\ \text{V}/\ 5.76\ \text{ohms}$

$\text{VA} = 7511\ \text{VA}$

$\text{VA} = 7511\ \text{VA}/1000$

$\text{VA} = 7.5\ \text{kVA}$

3. The correct answer is D.

38

What is the current of a 5-horsepower, 230-volt, 3-phase motor with efficiency of 80 percent and operating at 80 percent power factor?

A. 14.6 amperes
B. 15.3 amperes
C. 16.1 amperes
D. 17.4 amperes

ANSWER ______________

PROCEDURE TO ESTABLISH ANSWER

The question is about determining the current for a motor including efficiency and power factor.

1. Apply the basic formula for horsepower.
   HP (horsepower)
   HP = E (volts) x I (amps) x 1.73 x EFF x PF/ 746 (watts/horsepower)

   Or

   I = *HP x 746/ E x 1.73 x EFF x PF*
   I = 5 horsepower motor x 746 VA/ 230 V x 1.73(square root of 3) x .80eff x .80pf
   I = 3,730/255 eff and pf
   I = 14.6 amperes

2. The correct answer is A.

39

What is the minimum size electrical metallic tubing required for six 3 AWG Type THWN, three 8 AWG Type THW, and two 10 AWG Type THW conductors?

A. Metric Designator 41 (1 ½ in.)
B. Metric Designator 53 (2 in.)
C. Metric Designator 63 (2 ½ in.)
D. Metric Designator 78 (3 in.)

ANSWER _______________

PROCEDURE TO ESTABLISH ANSWER

The question is about calculating the required minimum size of EMT.

1. In Index find "Electrical metallic tubing," under which find "Number of conductors in, 358.22."

2. Section 358.22 refers to Table 1, Chapter 9 under which Note 6 for combinations of conductors refers to actual dimensions of the conductors or Tables 5 and 5A for dimensions of conductors and Table 4 for applicable size of tubing. Since no actual dimensions were provided in the question we will use the tables.

3. Referring to Table 5 for conductor areas,

   3 AWG THWN
   0.0973 cm x 6 conductors = 0.5838 sq. in.

   8 AWG THW
   0.0437 cm x 3 conductors = 0.1311 sq. in.

   10 AWG THW
   0.0243 cm x 2 conductors = 0.0486 sq. in.

   Total sq. in. of all conductors = 0.7635 sq. in.

4. Referring to Table 4 "Electrical Metallic Tubing" under "Over 2 Wires 40%" column:
   Metric Designator 35 (1¼ in.) tubing = 0.598 sq. in.
   Metric Designator 41 (1½ in.) tubing = 0.814 sq. in.

5. The correct answer is A.

40

A transformer has a 480-volt, 2-wire primary and a 120-volt, 2-wire secondary and will provide 20 amperes at 120 volts. When the secondary current is 20 amperes, how many amperes of current will be present in the primary?

A. 5 amperes
B. 10 amperes
C. 20 amperes
D. 40 amperes

ANSWER ______________

PROCEDURE TO ESTABLISH ANSWER

The question is about calculating transformer currents.

1. I_p = *(Primary Current)*
   E_s = *(Secondary Voltage)*
   I_s = *(Secondary Current)*
   E_p = *(Primary Voltage)*

   Therefore,

   $I_p = E_s \times I_s / E_p$

   I_p = 120 volts x 20 amps/ 480 volts

   I_p = 5 amperes

2. The correct answer is A.

41

An industrial heating unit is operated from a 3-phase, 240-volt system. The loads are connected in a delta configuration and balanced to provide an equal load on each phase. If the total load is 30 kVA, what is the current in the conductors supplying the unit?

A. 24 amperes
B. 42 amperes
C. 72 amperes
D. 125 amperes

ANSWER ______________

PROCEDURE TO ESTABLISH ANSWER

The question is about calculating current to a 3-phase load.

1. Using the basic formula:

$I = VA / 1.73 \times E$

$I = 30 \text{ kVA} \times 1000 / 240 \text{ volts} \times 1.73$

$I = 30{,}000 \text{ VA} / 415.2 \text{ V}_{eff}$

$I = 72.25$ amperes

2. The correct answer is C.

42

What is the minimum size Type THW copper conductor required to serve an ac transformer arc welder with a 60 percent duty cycle and at 43 amperes primary current? Note: The termination points (lugs) at the welder are rated at 60°C.

A. 10 AWG
B. 8 AWG
C. 6 AWG
D. 4 AWG

ANSWER ______________

PROCEDURE TO ESTABLISH ANSWER

The question is about the minimum size Type THW copper conductor required to serve an ac transformer welder.

1. In Index, find "Conductors," under which find "Welders, electric," under which find "Ampacity of supply, 630.11, 630.31," of which 630.11 is applicable.

2. Section 630.11(A) requires the ampacity of the supply conductors to be not less than current value determined by multiplying welder nameplate primary rating and the duty cycle factor per Table 630.11(A).

3. 43 amps x .78 (multiplier for 60 percent duty cycle)
   - 33.54 A.

4. In Index, find "Ampacities," under which find "Conductors, 310.14 through 310.21, Tables 310.16 through 310.21 , etc."

5. Since the lowest temperature rating of any connected termination, conductor, or device is 60°C, the 60°C column in Table 310.16 would be required to be used, instead of the 75°C column for Type THW.

6. In Table 310.16 , the next ampacity greater than 33.54 A would be 40 ampere for an 8 AWG copper conductor based on the temperature limitations on the terminals.

7. The correct answer is B.

   Note: Refer to Section 110.14(C)(1)(a) for information on temperature limitations for conductors and terminals.

43

What is the minimum volume required for a box containing three 12 AWG and three 10 AWG conductors?

A. 172 cm^3 (10.50 cu. in.)
B. 205 cm^3 (12.50 cu. in.)
C. 234 cm^3 (14.25 cu. in.)
D. 238 cm^3 (14.50 cu. in.)

ANSWER ______________

PROCEDURE TO ESTABLISH ANSWER

The question is about the minimum "cubic inch box volume" or "conductor fill" required for a specified number of conductors of different sizes.

1. The most direct pathway to the answer will be "boxes" and "conductors, number in box." In the Index find Boxes, (outlet, device, pull, and junction" "Conductors, number in box "314.16.

2. Section 314.16 requires the volume of the box to be not less than the fill calculation based on Section 314.16(B), which requires volumes to be added together.

3. Table 314.16(B) requires 2.25 cu. in. (36.9 cm^3) for 12 AWG and 2.50 cu. in. (41.0 cm^3) for 10 AWG.
   3 No. 12 AWG Conductors x2.25 cu. in. (36.9 cm^3)
   = 6.75 cu. in. (110.7 cm^3)

   3 No. 10 AWG x 2.50 cu. in. (41.0 cm^3)
   = 7.5 cu. in. (123.0 cm^3)

   Total = 14.25 cu. in. (233.7 cm^3)

4. The correct answer is C.

   Note: The question did not reference such things as clamps, devices, etc.

44

What is the minimum ampacity required for feeder conductors supplying two 25-horsepower and one 10-horsepower, 208-volt, 3-phase motors? (**N**ote: Motors are rated for continuous duty.)

A. 164 amperes
B. 181 amperes
C. 199 amperes
D. 218 amperes

ANSWER ______________

PROCEDURE TO ESTABLISH ANSWER

The question is about minimum ampacity for motor circuit conductors.

1. In Index, find "Motors," under which find "Conductors" under which find "Several motors, 430.24"; or in

Index find "Conductors" under which find "Motor circuits, 430-II."

2. Section 430.24 requires conductors to have an ampacity equal to the sum of the full-load current of all the motors plus 25 percent of the largest motor.

3. In Index, find "Tables," under which find "Motors," under which find "Full-load currents, Tables 430.247 through 430.250."

4. Table 430.250 provides the following full-load currents under the 208-volt column:
   FLC for 25-horsepower motor = 74.8 amps. (Remember, there are two 25 HP motors).

   FLC for 10-horsepower motor = 30.8 amps.

5. Minimum feeder ampacity:
   74.8 amps + 74.8 amps + 30.8 amps + 18.7 [25% of 74.8 (the largest motor)] = 199.1 amperes

6. The correct answer is C.

   Note: Either one of the two 25 HP motors may be chosen as the largest motor for the purposes of calculations.

45

A Schedule 40 rigid polyvinyl chloride (PVC) conduit contains two 500-kcmil Type THW conductors, one 250-kcmil Type THW conductor, and one 1 AWG bare conductor. The minimum permitted conduit size is:

A. Metric Designator 41 (1½ in.)
B. Metric Designator 53 (2 in.)
C. Metric Designator 63 (2½ in.)
D. Metric Designator 78 (3 in.)

ANSWER ______________

PROCEDURE TO ESTABLISH ANSWER

The question is about minimum Schedule 40 rigid polyvinyl chloride (PVC) conduit size required for a specified number of conductors of different sizes.

1. In Index, find "Conductor fill," under which find "Rigid polyvinyl chloride (PVC) conduit, 352.22."

2. Section 352.22 refers to Table 1, Chapter 9 under which Note 6 for combinations of conductors refers to the actual dimensions of the conductors or Tables 5 and 5A for dimensions of conductors and to Table 4 for applicable size of conduit. Since there aren't any actual dimensions provided, we will refer to the tables.

3. Referring to Table 5 for insulated conductor areas and Table 8 for bare conductor area:

   Two 500 kcmil Type THW (Table 5)
   0.7901 in^2 x 2 conductors = 1.5802 in^2

   One 250 kcmil Type THW (Table 5)
   0.4596 in^2 x 1 conductor = 0.4596 in^2

One 1 AWG bare conductor (Table 8)

0.0870 in^2 x 1 conductor = 0.0870 in^2

Total conductor area

$1.5802 \text{ in}^2 + 0.4596 \text{ in}^2 + 0.0870 \text{ in}^2 = 2.1268 \text{ in}^2$

4. Referring to Table 4 "Rigid PVC Conduit Schedule 40," under "Over 2 Wires 40%" column:

   Metric Designator 63 (2½ in.) = 1.878 sq. in.

   Metric Designator 78 (3 in.) = 2.907 sq. in.

5. The correct answer is D.

46 What is the minimum conductor ampacity required for a 460-volt, 3-phase feeder supplying one 10-horsepower squirrel cage induction motor, one 20-horsepower wound-rotor induction motor and one 30-horsepower wound-rotor induction motor?

A. 81 amperes
B. 91 amperes
C. 101 amperes
D. 177 amperes

ANSWER ______________

PROCEDURE TO ESTABLISH ANSWER

The question is about feeder conductor ampacity for several motors.

1. In Index find "Motors," under which find "Conductors," under which find "Several motors, 430.24." There is no specific subtitle in Index under "Motors" for feeder conductors.

2. Section 430.24 requires conductors to have an ampacity equal to the sum of the full-load current of all motors plus 25 percent of the largest motor.

3. In Index, find "Tables" under which find "Motors" under which find "Full-load currents, Tables 430.247 through 430.250."

4. Table 430.250 provides the applicable full-load currents for 3-phase motors. Using the "460 Volts" column find the following:

   FLC for 10-HP squirrel cage = 14 amperes

   FLC for 20-HP wound rotor = 27 amperes

   FLC for 30-HP wound rotor = 40 amperes

   25% of 30-HP FLC = 10 amperes (largest motor per 430.24)

Minimum feeder ampacity = 91 amperes

5. The correct answer is B.

47

A commercial building has a continuous lighting load of 41,250 volt-amperes at 240 volts, single phase. What is the minimum size Type THW copper feeder conductors permitted to be used?

A. 1/0 AWG
B. 2/0 AWG
C. 3/0 AWG
D. 4/0 AWG

ANSWER ______________

PROCEDURE TO ESTABLISH ANSWER

The question is about feeder conductor size.

1. Use the basic formula
   I = *P/E*
   I = 41,250 VA/ 240 volts
   I = 171.9 amperes

2. In Index, find "Feeders," under which find "Calculation of loads see Loads."

3. In the Index find "Loads" under which find "Feeders calculations of, 215.2(A), Art. 220, Annex D."

4. Section 215.2(A)(1)(a) requires feeder conductors to have sufficient ampacity to supply the load served and shall have an allowable ampacity not less than the noncontinuous load plus 125 percent of the continuous load.

5. Section 215.3 requires that, where the feeder supplies continuous loads, the overcurrent device shall not be less than 125 percent of a continuous feeder load, and the minimum feeder conductor shall have an ampacity equal to or greater than 125 percent of a continuous load, in this case:
   171.9 amperes x 125% = 214.8 amperes.

6. In Index, find "Conductors," under which find "Ampacities , 310.14 through 310.21 , Tables 310.16 through 310.21 , etc." Table 310.16 provides that a 4/0 AWG Type THW conductor has an ampacity of 230 amperes.

7. The correct answer is D.

   Note: Section 240.4 requires conductors to be protected in accordance with their ampacity. A 250-ampere overcurrent protective device protects the 4/0 AWG conductor and complies with the 125 percent requirement of Section 215.3 and 240.4(B) and (C).

48

What demand factor is permitted to be used for feeder load calculations for the first 20,000 volt-amperes of a hotel lighting load?

A. 60%
B. 70%
C. 80%
D. 100%

ANSWER ______________

PROCEDURE TO ESTABLISH ANSWER

The question is about feeder conductor demand factors for a hotel.

1. In Index find "Feeders," under which find "Calculation of loads, see Loads"

2. In the Index find "Loads" under which find "Feeders calculations for , 215.2(A), Art. 220, Annex D."

3. Scan Article 220 Part III and find Section 220.42, which refers to Table 220.42 for demand factors for that portion of the total branch-circuit load calculated for general lighting.

4. Table 220.42 permits the first 20,000 volt-amperes to have a 60 percent demand factor.

5. The correct answer is A.

49

What is the minimum required horsepower rating of a disconnect switch serving a multi-motor piece of machinery consisting of one 10-horsepower, one 20-horsepower, and one 30-horsepower motor, all of the squirrel-cage type rated at 3-phase, 460-volts? (**N**ote: Assume Design B Motors)

A. 50 HP
B. 60 HP
C. 75 HP
D. 100 HP

ANSWER ______________

PROCEDURE TO ESTABLISH ANSWER

The question is about the rating of motor disconnecting means.

1. In Index find "Motors," under which find "Disconnecting means, 430.75, 430.95, 430–Part IX, etc." You can also use the Table of Contents under which find "Article 430" under which find "Part IX, Disconnecting means," which will take you to Part IX.

2. Scan Article 430 Part IX and find Section 430.110, "Ampere Rating and Interrupting Capacity."

3. Section 430.110(C)(1), paragraph 2 states that the full-load current equivalent to the horsepower rating of each motor shall be selected from Tables 430.247, 430.248, 430.249 or 430.250 and shall be added to obtain an equivalent full-load current for the combined load.

4. For 3-phase motors, use Table 430.250 under the heading "Induction Type, Squirrel Cage and Wound-Ro-

tor Amperes:"

10-horsepower	=	14 amperes (FLA)
20-horsepower	=	27 amperes (FLA)
30-horsepower	=	40 amperes (FLA)
Total FLA	=	81 amperes (FLA)

5. Section 430.110(A) requires the disconnecting means to have an ampere rating of at least 115 percent of the full-load current rating:

   81 amperes (FLA) x 115%
   = 93 amperes

6. Since there is no motor shown in Table 430.250 that has a FLA of 93 amperes, go to the next higher FLA, which is the same as though there was one 75-horsepower motor, which shows 96 amperes.

7. Next, according to Section 430.110(C)(1) third paragraph, use the same method using the locked-rotor current selected from Table 430.251(B) (assume Design B motors). The values selected for the three motors are:

10-horsepower	=	81 amperes (LRC)
20-horsepower	=	145 amperes (LRC)
30-horsepower	=	218 amperes (LRC)
Total LRC	=	444 amperes (LRC)

8. Section 430.110(C)(2) requires the ampere rating of the disconnecting means to be not less than 115 percent of the sum of all currents at the full-load condition determined in accordance with 430.110(C)(1):

   444 amperes (LRC) x 1.15 = 510.6 amperes (LRC)

   Again, using the next higher value find a 75-horsepower motor from Table 430.251(B) which is 543 amperes.

9. The correct answer is C.

50 What is the recommended maximum voltage drop for feeders and branch circuits?

A. 1.5%
B. 3%
C. 5%
D. 6%

ANSWER ______________

PROCEDURE TO ESTABLISH ANSWER

The question is about the recommended maximum voltage drop for a feeder.

1. In Index find "Voltage and volts," under which find "Drop" under which find "210.19(A) Informational Note No. 3."

2. Section 210.19(A) Informational Note No. 3 states that a drop of 3 percent for branch circuits and 5 percent overall for feeders and branch circuits is recommended with this measurement being taken at the farthest outlet.

3. The correct answer is C.

51

What is the total calculated load of an installation of 12 m (40 ft) of a fixed multioutlet assembly, installed in a continuous length in an office occupancy?

A. 12 amperes
B. 16 amperes
C. 20 amperes
D. 30 amperes

ANSWER ______________

PROCEDURE TO ESTABLISH ANSWER

The question is about calculated load for multioutlet assemblies.

1. In Index find "Multioutlet assembly," under which find "Calculation of load, 220.14(H)."

2. Section 220.14(H) provides that each 1.5 m (5 ft) or fraction thereof of each separate and continuous length shall be considered as one outlet of not less than 180 VA capacity.

3. Proceed with the calculations:
   = 40 ft ÷ 5 ft
   = 8 calculated receptacles (Calculated at 180 VA each)

   8 calculated receptacles x 180 VA
   = 1440 VA
   = 1440 VA ÷ 120 V
   = 12 amperes

4. The correct answer is A.

52

A 3-phase, 4-wire, 480Y/277-volt feeder has all loads connected line to neutral. What is the current in neutral conductor? See sketch for loads.

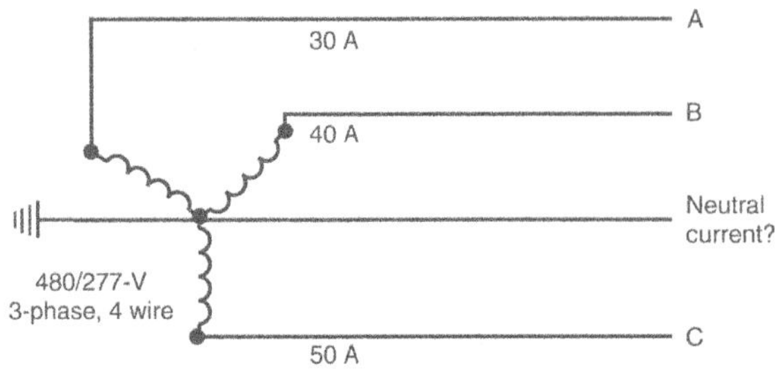

A. 10.00 amperes
B. 17.3 amperes
C. 20.00 amperes
D. 24.68 amperes

ANSWER _______________

PROCEDURE TO ESTABLISH ANSWER

The question is about neutral current in an unbalanced 3-phase circuit.

1. Answer Formula
   I of N
   = I2A + I2B + I2C - (IA x IB) - (IB x IC) - (IC x IA) / 1.73%

   Where: A = 30 amps
   B = 40 amps
   C = 50 amps

30 amps x 30 amps = 900 amps	30 amps x 40 amps = 1200 amps
40 amps x 40 amps = 1600 amps	40 amps x 40 amps = 2000 amps
50 amps x 50 amps = 2500 amps	30 amps x 50 amps = 1500 amps
Total = 5000 amps	Total = 4700 amps

   5000 amps - 4700 amps = 300 amps
   300 amps / 1.73% or √ 300 (square root of 300) = 17.32 amps

2. The correct answer is B.

53 What is the full-load current of a 208-volt, 3-phase, 1-horsepower squirrel cage motor?

A. 2.1 amperes
B. 4.2 amperes
C. 4.6 amperes
D. 4.8 amperes

ANSWER _______________

PROCEDURE TO ESTABLISH ANSWER

The question is about motor full-load current.

1. In Index find "Motors," under which find "Current, full load, see "Full-load current motors"

2. Under "Full-load current motors," find "Alternating current, Three-phase, Table 430.250."

3. In Table 430.250 find a 1-horsepower, 208-volt motor = 4.6 amperes.

4. The correct answer is C.

54

Determine the voltage a percolator would receive if the neutral conductor were broken on a multiwire circuit with a 720-watt percolator on Phase A and a 1200-watt fry pan on Phase B.

A. 120 volts
B. 150 volts
C. 240 volts
D. 720 volts

ANSWER ______________

PROCEDURE TO ESTABLISH ANSWER

The question is about voltages on multiwire circuits when the neutral is broken.

1. Find the resistance of each load:

   $R = E^2/P$
   R_A = (720 volt-amperes)
   R = E^2/VA
   R = 120^2/720 VA
   R = 14,400V^2/720 VA
   R = 20 ohms

   R_B = (1,200 volt-amperes)
   R = E^2/VA_B
   R = 120^2/1200 VA
   R = 14,400 V^2/1200 VA
   R = 12 ohms

2. Find total circuit resistance:

   20 ohms + 12 ohms = 32 ohms

3. Find the total series circuit current.

   I (amps) = E (volts)/ R (ohms)
   I = 240 V/32 ohms
   I = 7.5 amperes

4. Find voltage across percolator resistance load:

   $E = I \times R$
   E = 7.5 amps x 20 ohms
   E = 150 volts

5. The correct answer is B.

55

A non-motor-operated appliance used in a restaurant is to be supplied by an individual branch circuit. The appliance draws 20 amperes at 120 volts. What is the minimum and maximum size branch circuit permitted?

A. 20 and 25 ampere
B. 25 and 30 ampere
C. 20 and 30 ampere
D. 25 and 40 ampere

ANSWER _______________

PROCEDURE TO ESTABLISH ANSWER

The question is about branch-circuit loads for appliances.

1. In Index find "Branch circuits," under which find "Appliances, 210.23, 422–Part II."

2. Scan Article 422 Part II and find Section 422.10(A), that requires the rating to be not less than the marked rating of the appliance or, in this case, a minimum of a 20-ampere circuit.

3. Since this section does not give a maximum size, scan further and find Section 422.11(E). This limits the overcurrent device rating for a single non-motor-operated appliance of over 13.3 amps to 150 percent of the appliance rating:

   20 amps x 150% = 30 amps.

4. The correct answer is C.

   Note: Refer to Section 422.11(E)(3), where 150 percent of nameplate does not correspond to a standard overcurrent device rating, the next higher shall be permitted to be used.

56 Where branch-circuit loads are expected to be in continuous operation for more than three hours, the rating of the branch-circuit overcurrent device must be not less than what percentage of the load?

A. 115%
B. 125%
C. 150%
D. 175%

ANSWER _______________

PROCEDURE TO ESTABLISH ANSWER

The question is about branch-circuit rating for continuous loads.

1. In Index find "Continuous load," under which find "Applications, 210.19(A), 210.20(A), 215.2(A)"

2. Since the question is related to "branch circuits," a reference in Article 210 would be the best choice to start with. Section 210.19(A)(1)(a) requires that the rating of the branch-circuit overcurrent device be not less than 125 percent of the continuous load.

3. The correct answer is B.

   Note: Section 210.20(A) provides the same requirement and is specific to the rating of the overcurrent device.

57 In calculation of general lighting loads by square foot area, not considering any exceptions, the figure used for a bank-type occupancy is how many volt-amperes per square foot?

A. 3 VA per sq. ft.
B. 2 VA per sq. ft.
C. 1 VA per sq. ft.
D. 1.3 VA per sq. ft.

ANSWER ______________

PROCEDURE TO ESTABLISH ANSWER

The question is about calculation of lighting loads.

1. In Index find "Lighting," under which find "Branch circuits, calculation of load, 220.12."
2. Section 220.12 requires a load of not less than that specified in Table 220.12.
3. Reading down table to superscript [d] note "Banks," are considered to be the same as "office" lighting loads and then across the Table finds the unit load is required to be 1.3 volt-amperes per square foot.
4. The correct answer is D.

58 In calculating general lighting loads by occupancies, not considering any exceptions, how many volt-amperes per square foot are required for an office building when the actual number of general-purpose receptacle outlets is unknown?

A. 4 ½
B. 3 ½
C. 3
D. 2.3

ANSWER ______________

PROCEDURE TO ESTABLISH ANSWER

The question is about calculation of lighting and receptacle loads for an office building.

1. In Index find "Lighting," under which find "Branch circuits, calculation of load, 220.12."
2. Section 220.12 requires a load of not less than that specified in Table 220.12.
3. Read down table to "Office Building," and across find the unit load is required to be 1.3 volt-amperes per square foot
4. In 220.14(K)(2), an additional unit load of 1 volt-ampere per square foot is permitted , where the actual number of general-purpose receptacles is unknown.
5. The correct answer is D.

Note: If the actual number of receptacles is unknown, an additional 1 VA per square foot must be added to the lighting load. Section 220.14(K) requires the receptacle load be the larger of the calculated loads under 220.14(I) [180 volt-ampere/yoke, etc., or 1 volt-ampere/ft^2]. Since the question did not provide a number of receptacles , we used the 1 volt-ampere/ft^2.

59 The demand factor permitted to be used for feeder load calculations for 12 electric cooking ranges in a commercial restaurant is:

A. 65%
B. 75%
C. 80%
D. 100%

ANSWER ______________

PROCEDURE TO ESTABLISH ANSWER

The question is about the calculation of loads for commercial ranges.

1. In Index find "Ranges," under which find "Branch circuits" and then "Calculation of load, 220.55, Table 220.55."

2. Table 220.55 only applies to dwelling units, but (Informational Note No. 2) following Section 220.55 refers to Table 220.56 for commercial cooking units.

3. Table 220.56 provides a demand factor of 65 percent that is permitted to be applied to "six or more" commercial cooking units.

4. The correct answer is A.

60 A new restaurant that includes gas appliances has the following loads: 200 kVA for all loads except heating and cooling; 40 kVA heating; and 75 kVA cooling. Using the optional method of calculation, what percent demand factor is permitted to be applied to the total load?

A. 50% (amount over 200 kVA) + 200 kVA
B. 50% (amount over 150 kVA) + 150 kVA
C. 80% (amount over 200 kVA) + 200 kVA
D. 90% (amount over 325 kVA) + 325 kVA

ANSWER ______________

PROCEDURE TO ESTABLISH ANSWER

The question is about calculation of loads for a new restaurant.

1. In Index find "Loads," under which find "Feeders calculations for, 215.2(A), Art. 220, Annex D."

2. Scan Article 220 to find Part IV, Optional Feeder and Service Load Calculations." In 220–Part IV, find 220.88, New Restaurants."

3. Table 220.88 specifies demand factors based on the total load and depending upon whether the restaurant is "All-Electric Restaurant" or "Not All-Electric Restaurant."

4. The note below the table requires that all electrical loads including both heating and cooling be included in calculating the load. The total load is:

Other loads	=	200 kVA
Heating loads	=	40 kVA
A/C loads	=	75 kVA
Total	=	315 kVA

5. For a restaurant that is not all electric with a total 315 kVA connected load, Table 220.88 provides a 50 percent demand factor for that portion of the load over 200 kVA plus the original 200 kVA:

315 kVA – 200 kVA	= 115 kVA
115 kVA x 50%	= 57.5 kVA
200 kVA + 57.5 kVA	= 257.5 kVA

6. The correct answer is A.

61

A 100 mm x 100 mm (4" x 4") metal wireway that is 1.2 m (4 ft) long contains twenty-four 12 AWG Type THWN conductors, sixteen 10 AWG Type XHHW conductors and nine 6 AWG Type THW conductors that all pass through the same cross-sectional area of the wireway. Assume all conductors are current-carrying and the ambient temperature is 75°C. The allowable ampacity of the 6 AWG Type THW conductors is:

A. 22.75 amperes
B. 52 amperes
C. 55.25 amperes
D. 65 amperes

ANSWER ______________

PROCEDURE TO ESTABLISH ANSWER

The question is about the allowable ampacity of conductors installed in a metal wireway.

1. In the Index find "Wireways, metal" under which find "Number of conductors, 376.22."

2. Section 376.22(B) states that the adjustment factors in 310.15(C)(1) shall be applied only where the number of current-carrying conductors exceeds 30 at any cross section of the wireway.

3. Section 310.15(C)(1) Adjustment Factors "requires Table 310.15(C)(1) be applied where there are more than three current-carrying conductors in a raceway or cable.

4. Since there are forty-nine current-carrying conductors in the wireway, Table 310.15(C)(1) requires an adjustment factor(multiplier) of 35 percent be applied to each current-carrying conductor.

5. The allowable ampacity of the 6 AWG conductors with Type THW insulation from Table 310.16 is:
   = 65 amperes x .35 multiplier
   = 22.75 amperes

6. The correct answer is A.

62 A 100 mm x 100 mm (4" x 4") metal wireway that is 1.2 m (4 ft) long contains twenty-four 12 AWG Type THWN conductors, sixteen 10 AWG Type XHHW conductors and nine 6 AWG Type THW conductors. Assume all conductors are current-carrying and the ambient temperature is 75°C. What is the total square inch area of these conductors and what is the maximum permitted percent fill of the cross-sectional areas of all contained conductors at any cross section of a metal wireway?

A. 1.3102 sq. in. and 20%
B. 1.3614 sq. in. and 20%
C. 1.3102 sq. in. and 40%
D. 1.3614 sq. in. and 40%

ANSWER ______________

PROCEDURE TO ESTABLISH ANSWER

The question is about the conductors fill and percent fill in a metal wireway.

1. In Index find "Conductor fill," under which find "Wireways, 376.22, 378.22."

2. Section 376.22(A) states that the sum of the cross-sectional areas of all contained conductors at any cross section of a wireway shall not exceed 20 percent of the interior cross-sectional area of the wireway.

3. To determine the total square inch of the conductors, refer to Table 1, Chapter 9 under which Note 6 for combinations of conductors refers to the actual size of the conductors or Tables 5 and 5A for dimensions of conductors. Proceed with calculations using the tables as no actual dimensions were provided:

12 AWG Type THWN conductors	0.0133 sq. in. area x (24 conductors) = 0.3192 sq. in. area
10 AWG Type XHHW conductors	0.0243 sq. in. area x (16 conductors) = 0.3888 sq. in. area
6 AWG Type THW conductors	0.0726 sq. in. area x (9 conductors) = 0.6534 sq. in. area
Total area of conductors	= 1.3614 sq. in.

4. The correct answer is B.

63 A single-phase electric furnace rated at 20 kVA at 480 volts is connected to a single-phase 240-volt system. What will the outlet rating of the furnace be?

A. 5.8 kVA
B. 6.9 kVA
C. 7.2 kVA
D. 5 kVA

ANSWER ______________

PROCEDURE TO ESTABLISH ANSWER

The question is about calculating the output of a 480-volt furnace operating at 240 volts.

1. Using the VA and voltage rating of the furnace determine the resistance of the furnace.

$R = E \times E/VA, \text{ or } R = E^2 / P$
R = 480 V x 480 V / 20,000 VA
R = 230,400 V/20,000 VA
R = 11.52 ohms

2. Using the resistance of the furnace and voltage of the circuit determine the VA outlet.

$VA = E \times E/R, \text{ or } P = E^2R$
VA = 240 V x 240 V/11.52 ohms
VA = 57,600V^2/11.52 ohms
VA = 5000 volt-amperes

3. The correct answer is D.

64

What is the minimum size Type THW copper conductor permitted to supply an AC transformer arc welder with a 70 percent duty cycle and rated at 46 amperes primary current? The AC transformer welder connection terminals have a temperature rating of 60°C.

A. 10 AWG
B. 8 AWG
C. 6 AWG
D. 4 AWG

ANSWER _______________

PROCEDURE TO ESTABLISH ANSWER

The question is about the minimum size Type THW copper conductor permitted to serve an AC transformer welder.

1. In Index find "Conductors," under which find "Welders, electric," under which find "Ampacity of supply, 630.11, 630.31." Section 630.11 applies since an AC transformer arc welder is specified in the question.

2. Section 630.11(A), Individual Welders, requires the ampacity of the supply conductors to be not less than current values determined by multiplying welder nameplate primary rating and the duty cycle factor.

3. 46 amps x.84 (Multiplier for 70 percent duty cycle) = 38.64 amperes.

4. In the index, find "Connections" under which find "Temperature limitations, 110.14(C), 110.40."

5. Section 110.14(C) states that conductor shall be selected and coordinated so as not to exceed the lowest temperature rating of any connected termination. Therefore, even though Type THW is rated for 75°C (see Table 310.16), the 60°C column must be used because of the lower temperature rating of the terminals of the welder. Section 110.40 does not apply, since it is intended for use on systems rated over 1000 volts, nominal.

6. In Index, find "Ampacities" under which find "Conductors, 310.15, Tables 310.16 through 310.21."

7. In Table 310.16, read down the 60°C column to next ampacity higher than 38.64 amperes and read left to the AWG column, which requires 8 AWG Type THW copper conductor.

8. The correct answer is B.

65 What is the minimum volume required for a junction box containing six 12 AWG and three 10 AWG conductors?

A. 172 cm^3 (10.50 cu. in)
B. 205 cm^3 (12.50 cu. in)
C. 344 cm^3 (21 cu. in)
D. 238 cm^3 (14.50 cu. in)

ANSWER _______________

PROCEDURE TO ESTABLISH ANSWER

The question is about the minimum "cubic inch box volume" required for a specified number of conductors of different sizes.

1. In Index find "Boxes," under which find "Conductors, number in box, 314.16."

2. Section 314.16 requires the volume of the box to be not less than the fill calculation based on Section 314.16(B), which requires volumes to be added together.

3. Section 314.16(B)(1) for conductor fill requires the fill to be computed using Table 314.16(B).

4. Table 314.16(B) requires 36.9 cm^3 (2.25 cu. in) for every 12 AWG and 41.0 cm^3 (2.5 cu. in) for every 10 AWG conductor in the box.

   6 conductors (12 AWG) x 36.9 cm^3 (2.25 cu. in)
   = 221.4 cm^3 (13.5 cu. in)

   3 conductors (10 AWG) x 41.0 cm^3 (2.5 cu. in)
   = 123 cm^3 (7.50 cu. in)

   TOTAL
   = 344.4 cm^3 (21 cu. in) as the minimum box volume required

5. The correct answer is C.

CHAPTER 4
MOTORS

1 Conductors supplying the following group of 3-phase, 230-volt induction motors: one 3-horsepower, two 5-horsepower and one 7½-horsepower, shall have a minimum ampacity of how many amperes?

A. 62.0
B. 65.3
C. 67.5
D. 77.5

ANSWER ______________

PROCEDURE TO ESTABLISH ANSWER

The question is about the ampacity of conductors serving several (four) motors.

1. In Index, find "Motors," under which, find "Conductors 430-II," under which find "Ampacities 430.6."

2. Section 430.6(A)(1) generally requires that, where the full load current (FLC) rating of a motor is to be used to determine the ampacity of conductors, etc., the values in Tables 430.247 through 430.250, including the notes, shall be used instead of the actual full load ampere (FLA) rating marked on the motor nameplate.

3. Table 430.250 is applicable and reading down the horsepower column to desired horsepower, and across to the 230 V column:

One 3-horsepower rated 9.6 FLC x 1 motor	= 9.6 amps
Two 5-horsepower rated 15.2 FLC x 2 motors	= 30.4 amps
One 7½-horsepower rated 22.0 FLC x 1 motor	= 22.0 amps
Total amperes	= 62.0 amps

4. In Index, find "Motors," under which, find "Conductors, 430– II," under which, find "Several motors, 430.24."

5. Section 430.24 requires conductors supplying several motors to have an ampacity equal to the sum of the full-load current (FLC) rating of all the motors (which is 62 Amperes) plus 25 percent of the highest rated motor in the group:
   62 A + (22A x 25% for largest motor)
   = 62 A + 5.5 amperes
   = 67.5 amperes

6. The correct answer is C.

2 What are the minimum size feeder conductors for the following 3-phase, 460-volt wound-rotor motors: one 40-horsepower, one 150-horsepower, and one 10-horsepower?

A. 300 kcmil Type THWN copper
B. 350 kcmil Type THWN copper
C. 300 kcmil Type THWN aluminum
D. 350 kcmil Type THWN aluminum

ANSWER _______________

PROCEDURE TO ESTABLISH ANSWER

The question is about the size of feeder conductors serving several wound-rotor motors.

1. In Index, find "Motors," under which, find "Conductors, 430– II," under which, find "Several motors, 430.24."

2. In the Index you will also find "Conductors" under which find "Wound-Rotor Secondary 430.24." However, since nothing there that deals specifically with feeder conductors, 430.24 applies.

3. In general, 430.6 (A)(1) requires the full-load current (FLC) values given in Tables 430.247 through 430.250 to be used to determine the ampacity of conductors, the ampacity rating of switches and the branch-circuit short-circuit and ground-fault protection instead of the actual full load ampere (FLA) rating marked on the motor nameplate.

   Find full-load current of motors in Table 430.250. Section 430.24 requires the largest motor in the group to be taken at 125% of the FLC plus the sum of the other motors in the group as follows:

40-horsepower FLC	= 52 amps
150-horsepower FLC	= 180 amps
10-horsepower FLC	= 14 amps
Total	= 246 amps

   246 amps + 45 amps (25% of 180 amperes from the 150 HP motor) = 291 amperes

4. Referring to Table 310.16 and 110.14(C)(1)(b) for conductor termination temperatures, we find that this 291-ampere feeder load requires a minimum of 350 kcmil Type THWN CU conductors to carry the load.

5. The correct answer is B.

3

What are the minimum size conductors necessary for a feeder supplying three 3-phase, 208-volt, squirrel-cage motors: one 50-horsepower, one 25-horsepower and one 15-horsepower. Use smallest permissible size conductor.

A. 300 kcmil Type THW
B. 350 kcmil Type THW
C. 400 kcmil Type THW
D. 500 kcmil Type TW

ANSWER ______________

PROCEDURE TO ESTABLISH ANSWER

The question is about the size of feeder conductors supplying several squirrel-cage motors.

1. Refer to questions 1 and 2 in this Chapter for procedure steps.
   Table 430.250

50-horsepower	= 143 amps (FLC) x 125% (for the largest motor in the group).
	= 178.75 amperes (FLC)
25-horsepower	= 74.8 amperes (FLC)
15-horsepower	= 46.20 amperes (FLC)
Total	= 299.75 amperes (FLC) (178.75 ampere + 74.8 amperes + 46.20 amperes)

   Note: Since there is only one large motor, we choose the option of adding the additional 25% to the calculation by simply multiplying the largest motor full-load amps by 125%.
2. Since the asks for the minimum conductor size, refer to Table 310.16 .
3. Assume copper conductors since the does not state otherwise (see *NEC* 110.5).
4. The smallest size conductors having ampacity greater than 299.75 amperes are 350 kcmil Type THW rated at 310 amperes.
5. The correct answer is B.

4

What is the total full-load current for two 20-horsepower, 3-phase, wound-rotor motors rated 208 volts?

A. 84 amperes
B. 92.4 amperes
C. 108 amperes
D. 118.8 amperes

ANSWER ______________

PROCEDURE TO ESTABLISH ANSWER

The question is about full-load currents of motors.

1. In Index, find "Motors," under which, find "Current, full load. *see* Full-load current motors."

2. In Index, under "F" find "Full-load current motors" under which find "Alternating current" under which find "Three-phase, Table 430.250."

   Note that more experienced users of the *NEC* could find "430 Motors, Motor Circuits and Controllers" in the Table of Contents under which you will find "Part XIV. Tables." You can also go to Figure 430.1 and note "Tables 430.247 through 430.251(B) Part XIV.

3. In Table 430.250, read down the horsepower column to 20-horsepower, then across to the 208-volt column and find that a 20-horsepower motor is rated at 59.4 amperes.

4. 59.4 amps x 2 motors = 118.8 amperes

5. The correct answer is D.

   Note: Since the only asked for the full-load current of the motors and not for the minimum conductor size, the additional 25 percent of the largest motor load was not added.

5 Which one of the following statements concerning motors installed in a grain elevator where combustible dust is in the air under normal operating conditions in quantities sufficient to produce an explosion is not true?

A. A Class I, Division 1 explosionproof motor is acceptable for use in this Class II grain elevator dusty area.
B. A motor in a Class II, Division 1 area is required to be approved for material Group G atmospheres.
C. The motor may be a totally enclosed pipe-ventilated type, if temperature requirements are met.
D. A motor identified for Class II, Division 1 locations is acceptable.

ANSWER ______________

PROCEDURE TO ESTABLISH ANSWER

The question is about motors installed in hazardous location.

1. In Index, find "Hazardous atmospheres," under which, find "Class II locations, 500.5(C)." In Index, also find "Hazardous (classified) locations, Classes I, II, and III, Divisions 1 and 2," under which, find "Material groups, 500.6."

2. Section 500.5(C)(1) defines the area in to be a Class II, Division 1 location as the dust is present in normal operations.

3. In Index, find "Explosive atmospheres…" which refers to "Hazardous (classified) locations."

4. Under "Hazardous classified locations" find "Class II, 500.5(C), Article 502" under which find "Motors and generators, 502.125"

5. Section 502.125(A)(1) states that motors must be identified for Class II, Division 1 locations or be of the totally enclosed pipe-ventilated type. Answers C & D are true.

6. 500.6(B)(3) requires that equipment be approved, not only for the class of location, but also for the specific material group of dust that will be present. This is considered as a Class II, Division I, Group G location. Answer B is true. The motor mentioned in Answer A is not rated for the location or for the group specified in the so answer A is not true.

7. The correct answer is A.

6 Which one of the following statements about a permanently wired optional standby system is true?

A. It is required to be listed for emergency use.
B. The optional standby equipment is not required to be suitable for available fault current.
C. The user of the standby system is permitted to select the connected load when manual transfer equipment is used.
D. The system shall be designed to automatically transfer the connected load upon failure of the normal system.

ANSWER ______________

PROCEDURE TO ESTABLISH ANSWER

The question is about a permanently wired optional standby systems.

1. In Index, find "Optional standby systems," under which, find "Capacity and rating, 702.4" and "Transfer equipment, 702.4(B), 702.5."

2. Where manual transfer equipment in used, 702.4(B) requires this type of a system to have adequate capacity and rating for the supply of all equipment intended to operate at one time. The equipment is required to be suitable for the maximum available fault current at its terminals based on 702.4(A), and the user is permitted to select the load connected to the system in 702.4(B)(1) last sentence. Answer B is not true and Answer C is true.

3. Section 702.5(D) only requires the transfer equipment to be designed to prevent inadvertent interconnection of normal and alternate sources but does not mandate automatic transfer. Answer D is not true.

4. As noted above, 702.5(D) states the transfer equipment must be suitable of the intended use and shall be listed, designed, and installed to prevent inadvertent interconnection but not for emergency use so answer A is not true.

5. The correct answer is C.

7 To determine the ampacity of conductors supplying a 10-horsepower, 3-phase, 208-volt squirrel cage motor with nameplate FLA of 29.5 amperes, which one of the following values is to be used?

A. 28 amperes
B. 29.5 amperes
C. 30.8 amperes
D. 32.2 amperes

ANSWER ______________

PROCEDURE TO ESTABLISH ANSWER

This is about sizing motor branch-circuit conductors. In Index, find "Motors," under which, find Conductors, 430– II" under which find "Ampacities 430.6.

1. In general, 430.6(A)(1) requires the full-load current values given in Tables 430.247 through 430.250 to be used to determine the ampacity of conductors, the full load current (FLC) rating of switches and the branch-circuit short-circuit and ground-fault protection instead of the actual full load ampere (FLA) rating marked on the motor nameplate.

2. According to Table 430.250, the FLC for a 10-horsepower motor is 30.8 amperes.

3. The correct answer is C.

8 A 3-horsepower hermetic refrigerant motor-compressor has a nameplate indicating only the following: 3-phase, 480-volts, 60-cycle, and code letter F. What is the minimum locked-rotor kVA of the motor?

A. 3.0 kVA
B. 3.75 kVA
C. 16.77 kVA
D. 18.0 kVA

ANSWER _______________

PROCEDURE TO ESTABLISH ANSWER

The question is about locked-rotor currents of motors.

1. In Index, find "Motors," under which, find "Code letters, Table 430.7(B)."

2. Section 430.7(B) states that code letters marked on motor nameplates to show motor input with locked rotor must comply with Table 430.7(B).

3. Code letter F indicates between 5 and 5.59 kVA per horsepower; therefore,
   3-hp x 5.59 (kVA per horsepower) = 16.77 kVA.

4. The correct answer is C.
   Note: Although this involves a motor-compressor, which is covered by Article 440, Section 440.3(A) states that the provisions of Article 430 apply except as modified in Article 440. Article 440 does not modify Section 430.7(B) or Table 430.7(B).

9 What size Type THW copper conductors are required to supply a 40-horsepower, 230-volt, 3-phase squirrel-cage motor? All equipment and terminations are listed for 75°C.

A. 1 AWG
B. 2 AWG
C. 3 AWG
D. 4 AWG

ANSWER _______________

PROCEDURE TO ESTABLISH ANSWER

The question is about sizing motor branch-circuit conductors. Determine the motor full-load current value from Table 430.250 — 40-horsepower motor
FLC = 104 amperes.

Note: In general, 430.6 requires the full-load current values given in Tables 430.247 through 430.250 to be used to determine the ampacity of conductors, the full load current (FLC) rating of switches and the branch-circuit short-circuit and ground-fault protection, instead of the actual full load ampere (FLA) rating marked on the motor nameplate.

1. In Index, find "Motors," under which, find "Conductors, 430– II," under which, find "Single motor, 430.22."

2. Section 430.22(A) requires the conductors to have an ampacity of not less than 125 percent of FLC rating. Therefore, 104 amperes x 125% = 130 amperes.

3. In Table 310.16 , Type THW insulation is listed in the 75°C column. [See 110.14(C)(1)(b)]. 1 AWG copper has an ampacity of 130 amperes.

4. The correct answer is A.

10

In determining the ampacity of a feeder supplying several continuous duty motors, which one of the following statements is true?

A. Size the feeder for 100 percent of the sum of full-load currents of all of the motors based on Tables 430.248 through 430.250.
B. Size the feeder for 100 percent of the sum of full-load currents of all of the motors based on the nameplate currents.
C. Size the feeder for 125 percent of the nameplate current of the motor with the highest full-load current, plus 100 percent of the full-load nameplate current of all other motors.
D. Size the feeder for 125 percent of the full-load current as listed in Tables 430.247 through 430.250 of the motor with the highest full-load current, plus 100 percent of the full-load current of all other motors, as listed in the tables.

ANSWER ______________

PROCEDURE TO ESTABLISH ANSWER

The question is about the circuit conductors for two or more motor branch circuits.

1. In Index, find "Motors," under which, find "Conductors, 430– II" under which find "Several motors, 430.24.

2. Section 430.24 requires the supply conductors to have an ampacity equal to the sum of the FLC rating of all the motors plus 125 percent of the highest rated motor in the group.

3. The correct answer is D.

Note: Section 430.6(A) requires the use of Tables 430.247 through 430.250 in calculating loads, overcurrent protection, etc., not the nameplate rating on the motor.

11

A single motor air-conditioning unit is rated at 32 amperes. What is the minimum ampacity required for the conductors supplying this unit?

A. 32 amperes
B. 40 amperes
C. 56 amperes
D. 60 amperes

ANSWER ______________

PROCEDURE TO ESTABLISH ANSWER

The question is about sizing of conductors for air-conditioning motors.

1. In Index, find "Air-conditioning and refrigerating equipment," under which, find "Branch circuit," under which, find "Conductors, 440– IV," under which, find "Single motor-compressor for, 440.32."
2. Section 440.32 requires that the branch-circuit conductors supplying a single motor-compressor have an ampacity not less than 125 percent of the motor-compressor rated-load current or the branch-circuit selection current, whichever is greater.
3. Since the branch-circuit selection current is not given, the rated-load current is used. Therefore, 125 percent of the rated-load current (32 amperes) is 125% x 32 amps = 40 amperes. The conductors must have an ampacity of at least 40 amperes.
4. The correct answer is B.

12

Which one of the following statements about grounding a 240-volt, 1-horsepower submersible water pump in a metal well casing is true?

A. It is permissible to use the metallic water pipe from the well as the equipment grounding conductor.
B. It is permissible to ground the motor solely to the well casing.
C. It is not required to be grounded since it is not subject to contact.
D. It shall be grounded by *Code* as a fixed motor.

ANSWER ______________

PROCEDURE TO ESTABLISH ANSWER

The question is about the grounding of a submersible water pump, which is considered as fixed equipment.

Note: fixed equipment is equipment that is fastened in place, per the Code, as opposed to portable or mobile equipment.In Index, find "Motors," under which, find "Grounding, 250–VI, 430.12(E), 430.96, 430– XIII."

1. Scan Article 430, Part XIII, and find "430.242, Stationary Motors," where subsections (1) and (4) require the pump to be grounded.
2. In the Index, also find "Equipment," under which find "Grounding, 250–Part VI." Scan Article 250 Part

VI, and find "250.112(L)," which requires motor-operated water pumps to be connected to an equipment grounding conductor. The method would be the same as for any fixed motor. It is further stated in 250.112(M) that the well casing shall be connected to the pump circuit equipment grounding conductor.

3. The correct answer is D.

13

A 3-phase, 40-horsepower, 460-volt, ac squirrel-cage induction motor without code letters and having full-voltage starting is protected against short circuits and ground faults by dual-element fuses. The absolute maximum standard rating permitted for the fuses is how many amperes?

A. 65
B. 90
C. 100
D. 110

ANSWER ______________

PROCEDURE TO ESTABLISH ANSWER

The question is about sizing motor branch-circuit, short-circuit and ground-fault protection.

1. In Index, find "Motors," under which, find "Full-load Current", see "F" for Full-load current motors," under which, find "Alternating current, Three-phase, Table 430-250."

2. Also in Index, under "Motors," find "Rating or setting of branch-circuit short-circuit and ground-fault devices, Table 430.52" and "Short circuit protection, 430– IV, 430– V, 430.130."

   Note that an experienced user of the *NEC* could look in the Table of Contents under Article 430 and find "Part IV, Motor Branch-Circuit Short-Circuit and Ground-Fault Protection, which will bring you directly to 430.52.

3. Scan Article 430, Part IV, and find "430.52, Rating or Setting for Individual Motor Circuit."

4. Section 430.52 refers to Table 430.52, which specifies 175 percent of full load current (FLC) as the maximum for dual-element fuses.

5. The full load current is 52 amperes from Table 430.250.
   52 A x 175% = 91 amperes.

6. A check of 240.6(A), Fuses and Fixed-Trip Circuit Breakers indicates the standard ampere ratings for fuses and inverse time circuit breakers shall be considered as shown in Table 240.6(A). Table 240.6(A) does not list 91 amperes as a standard overcurrent device rating. Section 430.52(C)(1), Exception No. 1, permits using the next higher standard size where the values for branch-circuit short-circuit and ground-fault protective devices determined by Table 430.52 do not correspond to the standard sizes.

7. The next higher standard size is a 100-ampere fuse.

8. Section 430.52(C)(1), Exception No. 2(b), indicates that where the 100-ampere fuse (allowed by Exception No. 1) is not sufficient for the starting current of the motor, the time-delay fuse may be increased to ot more than 225 percent of the full-load current (FLC). 52 A x 225% = 117 amps.

9. Since there is no permission to round up to the next standard size (Remember, the question stated "The

absolute maximum standard rating" not the rounded-up value. Trick question!!), the next lower size is selected. The absolute maximum standard fuse would be 110 amps. Section 430.52 Exception No. 2(b) specifies there is no case where the 225% of the full-load current can be exceeded.

10. The correct answer is D.

14

A 30-horsepower, 3-phase, 230-volt squirrel-cage Design B energy efficient motor has branch-circuit protection provided by an instantaneous trip circuit breaker, which has a listed controller for an instantaneous trip circuit breaker. The initial setting of the instantaneous trip circuit breaker is:

A. 800 amperes
B. 900 amperes
C. 1000 amperes
D. 1200 amperes

ANSWER ______________

PROCEDURE TO ESTABLISH ANSWER

The question is about motor branch-circuit overcurrent protection for a Design B energy efficient motor. As noted in previous questions, refer to Table 430.250 for motor full-load currents and to Section 430.52(C) and Table 430.52 for branch-circuit protection requirements.

1. Table 430.250 lists a 30-horsepower, 3-phase, 230-volt motor FLC at 80 amperes.

2. Table 430.52 lists the instantaneous trip breaker for a Design B energy efficient motor at 1100 percent. Therefore, 80 A x 1100% = 880 amperes.

3. Since 880 amperes is not a standard size per 240.6(A) and Table 240.6(A), 430.52 Exception No. 1 would permit the next higher size, rating or possible setting for an adjustable trip breaker that does not exceed the higher standard ampere rating. Note there is no 800-ampere limitation in 430.52 Exception No. 1, as there is in 240.4(B) for general applications, so the next higher standard ampere rating of 1000 amperes is permitted.

4. The correct answer is C.

15

At construction sites, 125-volt, single-phase, 20-ampere receptacle outlets mounted on portable generators for temporary wiring that were manufactured after January 1, 2011 shall have GFCI protection, where the generator rated capacity does not exceed which of the following:

A. 10 kW
B. 8 kW
C. 5 kW
D. 15 kW

ANSWER ______________

PROCEDURE TO ESTABLISH ANSWER

The question is about GFCI protection for receptacles on portable generators used on construction sites.

1. In Index, find "Temporary installations, Art. 590," under which, find "Receptacles, 590.4(D), 590.6."

2. Section 590.6(A)(3) requires all 125-volt, single-phase, 15-, 20-, and 30-ampere receptacle outlets that are part of a 15 kW or smaller generator to have listed ground-fault circuit-interrupter protection for personnel.

   Listed cord sets or devices incorporating listed ground-fault circuit-interrupter protection for personnel identified for portable use shall be permitted for use with 15- kW or less portable generators manufactured or remanufactured prior to January 1, 2011. Since the question states the generator was manufactured after January 1, 2011 the provision requiring GFCI protection is applicable.

3. The correct answer is D.

16

Branch-circuit conductors supplying a single motor shall have an ampacity not less than what percent of the motor full-load current (FLC) rating?

A. 100
B. 115
C. 125
D. 175

ANSWER ______________

PROCEDURE TO ESTABLISH ANSWER

The question is about ampacity of motor branch-circuit conductors.

1. In Index, find "Motors," under which, find "Conductors, 430– II" under which find "Single motors 430.22."

2. Section 430.22(A) states the ampacity of a branch-circuit conductor supplying a single motor shall not be less than 125 percent of the full-load current (FLC) rating of the motor.

3. The correct answer is C.

17

What is the maximum length of flexible metal conduit permitted to be used for a raceway connection to a 2400-volt motor terminal enclosure?

A. 900 mm (3 ft)
B. 1.8 m (6 ft)
C. 2.5 m (8 ft)
D. 3.0 m (10 ft)

ANSWER ______________

PROCEDURE TO ESTABLISH ANSWER

The is about the use of flexible metal conduit as a raceway to a 2400-volt motor.

1. In Index, find "Flexible metal conduit (Type FMC)," under which, find "Uses permitted 348.10" and "Uses not permitted, 348.12."

2. Sections 348.10 and 348.12 do not place any length restrictions on this type of raceway. However, there may be a limit to its use on high voltage motors.

3. In Index, find "Motors," under which, find "Over 1000 volts, 430–XI."

4. Scan Article 430 Part XI and find "430.223," which permits the use of flexible metal conduit or liquidtight flexible conduit not exceeding 1.8 m (6 ft) in length.

5. The correct answer is B.

18 A 15-horsepower squirrel-cage induction motor (without brushes, switching mechanism or similar arc-producing devices), used outdoors to pump gasoline at a bulk storage plant, must the motor always have a label suitable for the following location:

A. Class I, Division 1
B. Class II, Division 1
C. Class I, Division 2
D. Class or Division marking is not required for squirrel-cage induction motors for this application.

ANSWER ______________

PROCEDURE TO ESTABLISH ANSWER

The question is about motors in a hazardous area, specifically a gasoline bulk plant.In Index, find "Hazardous (classified) locations, Classes I, II, and III, Divisions 1 and 2," under which find "Bulk storage plants, 515.3."

1. Section 515.3 refers to Table 515.3, which establishes under "Pumps, bleeders, withdrawal fittings — Outdoors" a Class I, Division 2 area within 900 mm (3 ft) extending in all directions, a height of 450 mm (18 in.) above grade level as well as horizontally within 10 feet from any edge of a pump located outdoors.

2. Having established the area classification, the next thing to check is the wiring and equipment that may be utilized in that area. Section 515.4 refers to Article 501.

3. Refer back to the Index and find "Hazardous (classified) locations, Class I," under which, find "Motors and Generators, 501.125 ,"

4. The last sentence of 501.125(B), regarding Class 1 Division 2 locations, permits the use of open or non-explosionproof motors, such as squirrel-cage induction type.

5. The correct answer is D.

19 Which one of the following statements regarding a generator set driven by a prime mover and used for a legally required standby system is true?

A. It shall be provided with a means to automatically start the prime mover on failure of normal service.
B. It shall be provided with on-site fuel supply of not less than four hours capacity if an internal combustion engine is used.
C. The prime mover may depend upon a municipal water supply for cooling.
D. The generator may provide the charging means for a battery used for starting the prime mover.

ANSWER ____________

PROCEDURE TO ESTABLISH ANSWER

The question is about standby generators for a legally required standby system.

1. In Index, find "Generators," under which, find "Standby systems, Art. 701, 702."
2. Article 701 concerns legally required standby systems. Since no other subtitles are listed in the Index, scan Article 701 for the specific answers referred to in the question.
3. Section 701.12(D) requires the prime mover to be started automatically. Answer A is true.
4. Section 701.12(D)(2) requires an on-site fuel supply for not less than two hours. Answer B is not true.
5. Section 701.12(D)(3) states that the prime mover shall not be solely dependent upon a municipal water supply for cooling. Answer C is not true.
6. Section 701.12(D)(4) states that a storage battery used for starting the prime mover shall have an automatic charging means independent of the generator set. Answer D is not true.
7. The correct answer is A.

20 If the disconnecting means for a 230-volt ac, 5-horsepower motor is a motor circuit switch, the motor circuit switch must be horsepower rated not less than what rating?

A. 5 horsepower
B. 7 horsepower
C. 10 horsepower
D. 15 horsepower

ANSWER ____________

PROCEDURE TO ESTABLISH ANSWER

The question is about disconnecting means for motors, specifically a switch.

1. In the index under "Motors under which, find " Disconnecting Means, 430.75, 430.95, 430-IX, etc.
2. Scan Article 430, Part IX and find "430.109, Type."
3. Section 430.109 requires a horsepower rated switch in 430.109(A), unless another type of disconnect is

otherwise permitted in (B) through (G), under the conditions specified.

4. Section 430.109(A)(1) requires a listed motor-circuit switch rated in horsepower.

5. 5 hp is the rated motor horsepower stated in the question.

6. The correct answer is A.

21 The minimum rating of a circuit breaker used as a motor disconnecting means is required to be at least what percentage of the full-load current rating of the motor?

A. 100%
B. 115%
C. 125%
D. 200%

ANSWER ______________

PROCEDURE TO ESTABLISH ANSWER

The question is about the ampere rating of a circuit breaker used as a motor disconnecting means.

1. In Index, find "Motors," under which, find "Disconnecting means, 430.75, 430.95, 430– IX, etc." or you can go to Figure 430.1 and find "Disconnecting means, 430.101 through 430.113."

2. Scan Article 430, Part IX, and find "430.110(A)," which requires the disconnecting means for motor circuits to be at least 115 percent of the full-load current rating of the motor.

3. The correct answer is B.

22 Which one of the following statements about the location of a single motor disconnecting means, controller, and motor operating at 1000 volts or less is true without using any exceptions?

A. The controller disconnecting means may be located out-of-sight of the controller.
B. A disconnecting means is always required in sight of the motor.
C. The controller disconnecting means is never required to be capable of being locked in the open position.
D. A disconnecting means is not required in sight of the motor, if the controller disconnecting means is individually capable of being locked in accordance with 110.25.

ANSWER ______________

PROCEDURE TO ESTABLISH ANSWER

The question is about motors and controllers being out-of-sight of disconnecting means.

1. In Index, find "In sight from," under which, find "Motors, motor controllers, 430.102."

2. Without using any exceptions, 430.102(A) requires a disconnecting means for each controller, and for it to be located in sight from the controller location. There are exceptions for motor controllers capable of being locked in the open position for motors operating over 1000 volts and valve actuator motor assemblies (VAM) neither of which are applicable in the of the asked. Answer A is not true.

3. Section 430.102(B) requires a disconnecting means to be located in sight from the motor unless the controller disconnecting means is used to disconnect the controller and the motor, and it is individually capable of being locked in accordance with 110.25. Answer D is true. Answers B and C are not true.

4. The correct answer is D.

23 A general-use switch is permitted as the disconnecting means for motors rated over 2 horsepower up to and including 100 horsepower when controlled by an autotransformer-type controller under all of the following conditions EXCEPT where:

A. The switch supplies a fire pump provided with overcurrent devices rated for locked rotor condition.
B. The switch supplies a generator provided with overload protection.
C. The controller is capable of interrupting the locked-rotor current of the motor.
D. The controller is provided with running overload protection not exceeding 125 percent of the motor full-load current rating.

ANSWER ______________

PROCEDURE TO ESTABLISH ANSWER

The question is about the type of disconnecting means permitted for motor circuits.

1. In Index, find "Motors," under which, find "Disconnecting means, 430.75, 430.95, 430– IX, etc." or you can go to Figure 430.1 and find "Disconnecting means, 430.101 through 430.113."

2. Scan Article 430, Part IX, and find "430.109(A)(1)," which requires horsepower-rated switches when used as a motor disconnecting means.

3. Section 430.109(D), however, permits a general-use switch for autotransformer-type controllers driving generators meeting the conditions stated in statements (1) is true in 430.109(D)(1) as answer B, (2) is true in 430.109(D)(1) as answer C and (3) is true in 430.109(D)(1) as answer C . No such exemption is stated for fire pumps so Answer A is not true.

4. The correct answer is A.

24 A ¼-horsepower, 115-volt continuous-duty automatically started motor with a nameplate full-load current (FLC) rating of 5.8 amperes and a service factor of 1.15 has a separate individual overload device. If the overload device is sufficient to permit starting the motor, its maximum permitted rating is how many amperes?

A. 6.67
B. 7.25
C. 10.0
D. 15

ANSWER ______________

PROCEDURE TO ESTABLISH ANSWER

The question is about motor overload protective device ratings.

1. In Index, find "Motors," under which, find "Overload protection, 430– III, etc." You can also go to the Table of Contents under "Article 430 Motors, Motor Circuits, and Controllers" and find "Part III. Motor and Branch-Circuit Overload Protection, Page 70-315 " which will take you to Part III. You can also go to Figure 430.1 and find "Motor Branch-Circuit Overload Protection" 430.31 through 430.44.

2. Scan Article 430 Part III and find "430.32(B), One Horsepower or Less, Automatically Started," in which (1) refers you back to 430.32(A)(1), which states that a separate overload device for a motor with a service factor of 1.15 or greater shall be rated at no more than 125 percent of the motor nameplate full-load current rating.

3. 5.8 (FLC) full load current x 125% = 7.25 amperes.

4. The correct answer is B.

25 The conductors from a generator to the first overcurrent device are not required to be sized larger than the nameplate rating of the generator under which one of the following conditions?

A. Where the conductors terminate in a single overcurrent device and are not over 7.5 m (25 ft) in length
B. Where the design and operation of the generator prevent overloading
C. Where the load supplied is all noncontinuous
D. Where the rating of the generator does not exceed the short-circuit rating of the device supplied

ANSWER ______________

PROCEDURE TO ESTABLISH ANSWER

The question is about the ampacity of conductors from a generator.

1. In Index, find "Generators," under which, find "Conductor, ampacity of, 445.13."

2. Section 445.13 requires conductors from the generator terminals to the first overcurrent device to be sized 115 percent of the nameplate current rating.

3. The exception permits the conductors to be not less than 100 percent of the nameplate current rating where the design and operation of the generator prevent overloading.

4. The correct answer is B.

26 Which one of the following statements regarding ratings and overload protection for generators is true?

A. Generators are rated by kilowatts or kilovolt-ampere output.
B. Generators are rated by horsepower output.
C. Constant-voltage generators do not require protection from overloads.
D. Circuit breakers are not acceptable for providing overload protection for constant-voltage generators.

ANSWER ______________

PROCEDURE TO ESTABLISH ANSWER

The question is about rating and overcurrent protection of generators.

1. In Index, find "Generators," under which, find "Marking, 445.11."
2. Section 445.11 provides that each generator shall be provided with a nameplate, etc., and the ratingin kilowatts or kilovolt amperes, etc. Answer A is true. Answer B is not true.
3. In Index, find "Generators," under which, find "Overcurrent protection, 445.12."
4. Section 445.12(A) provides that constant-voltage generators shall be protected from overloads by protective means including circuit breakers. Answers C and D are not true.
5. The correct answer is A.

27

Where generators are used to provide alternate power for a hospital essential electrical system, the transfer time from normal power to generator power for the life-safety branch shall be a maximum of:

A. 5 seconds
B. 10 seconds
C. 15 seconds
D. 20 seconds

ANSWER ____________

PROCEDURE TO ESTABLISH ANSWER

The question is about generator connection or transfer time for a hospital life safety branch.

1. In Index, find "Health care facilities," under which, find "Essential Electrical Systems, 517– III," under which, find "Connections," under which is listed "Life safety branch, 517.43."
2. Although 517.43 specifically addresses the life safety branch and the equipment branch and allows a maximum of 10 seconds to restore operation of the life safety branch.
3. The correct answer is B.

28

A 2-wire dc generator may have an overcurrent device in any of the following EXCEPT:

A. Armature leads
B. Series circuit
C. Compound winding on the armature
D. Shunt field conductors

ANSWER ____________

PROCEDURE TO ESTABLISH ANSWER

The question is about overcurrent protection for 2-wire dc generators.

1. In Index, find "Generators," under which, find "Overcurrent protection, 445.12."
2. Section 445.12(B) states the overcurrent device shall not open in the shunt field.
3. The correct answer is D.

29 What is the minimum ampacity of the phase conductors between the terminals of a 100-ampere rated generator, which is not designed to prevent overloading, to the first overcurrent device?

A. 100
B. 115
C. 125
D. 150

ANSWER _______________

PROCEDURE TO ESTABLISH ANSWER

The question is about the ampacity of conductors from the generator to the first overcurrent device.

1. In Index, find "Generators," under which, find "Conductor, ampacity of, 445.13."
2. Section 445.13 requires the ampacity of the phase conductors from the generator terminals to the first overcurrent device to be not less than 115 percent of the nameplate current rating of the generator.
3. The exception to Section 445.13 indicates that it does not apply to this installation.
4. 100 amperes x 115% = 115 amperes.
5. The correct answer is B.

30 Which one of the following statements about a generator used for a legally required standby system is NOT true?

A. Means shall be provided to automatically start the prime mover.
B. Means shall be provided for automatic transfer of the required circuits.
C. A storage battery used for starting the prime mover shall be equipped with an automatic charging means independent of the generator.
D. A storage battery used for starting the prime mover may be equipped with a charging means supplied from the generator.

ANSWER _______________

PROCEDURE TO ESTABLISH ANSWER

The question is about generators used for legally required standby systems.

1. In Index, find "Legally required standby systems," under which, find "Sources of power, 701– III" and "Transfer Equipment, 701.5." Scan Article 701, Part III, and find "701.12(D) , Generator Set."
2. Section 701.12(D)(1) requires means to be provided for automatically starting the prime mover and for automatic transfer. Answer A is true.
3. Section 701.12(D)(4) requires the storage battery used for starting the prime mover to be equipped with a charging means independent of the generator. Answer C is true. Answer D is not true.
4. Section 701.5(A) requires transfer equipment to be automatic. Answer B is true.
5. The correct answer is D.

31

A direct-current generator neutral conductor that must carry ground-fault currents shall not be smaller than what percentage of the size of the largest phase conductor?

A. The size established in Section 250.30(A)
B. 80%
C. 100%
D. 125%

ANSWER ______________

PROCEDURE TO ESTABLISH ANSWER

The question is about sizing neutral conductors for generators.

1. In Index, find "Generators," under which, find "Conductor, ampacity of, 445.13."

2. Section 445.13(A) requires the neutral conductor to be not smaller than required by 250.30(A).

3. The correct answer is A.

32

Which one of the following statements about generator overcurrent protection is true?

A. Two-wire dc generators may have overcurrent protection in one wire only as longas the overcurrent device is actuated by the entire current generated and the overcurrent device opens both the armature and shunt field leads.
B. Generators operating at 65 volts or less are considered as protected by the motor overcurrent protective device if they operate when the generator is delivering no more than 200 percent rated full-load current.
C. Two-wire, direct-current generators always require overcurrent protection in both conductors.
D. Two-wire, direct-current generators used to obtain neutrals for 3-wire systems require overcurrent devices that will disconnect the 3-wire system in case of unbalances of voltage or current.

ANSWER ______________

PROCEDURE TO ESTABLISH ANSWER

The question is about overcurrent protection for generators.

1. In Index, find "Generators," under which, find "Overcurrent protection 445.12."

2. Section 445.12(B) does not permit the overcurrent device to open the shunt field and only requires overcurrent protection in one conductor under certain conditions. Answer A is not true.

3. Section 445.12(C) provides that the generator is considered protected where the overcurrent device of the driving motor will operate when the generator output is not in excess of 150 percent of its full-load current rating. Answer B states 200% and is not true.

4. Section 445.12(B) permits two-wire, dc generators to have overcurrent protection in only one conductor. Answer C is not true.

5. Section 445.12(D) requires two-wire, dc generators used to obtain neutral points for 3-wire systems to be equipped with overcurrent devices to disconnect the system in case of unbalance. Answer D is true.

1. The correct answer is D.

33

Generators covered by Article 445 shall be located in accordance with all of the following EXCEPT:

A. Open generators having spark-producing contacts shall not be installed on wooden floors.
B. Open generators having spark-producing contacts shall be located so that sparks cannot reach adjacent combustible material.
C. So that adequate ventilation is provided
D. So that lubrication of bearings can be readily accomplished

ANSWER ______________

PROCEDURE TO ESTABLISH ANSWER

The question is about the location of generators.

1. In Index, find "Generators," under which, find "Location, 445.10."

2. Section 445.10 does not specifically address the statements above but does require generators to meet the requirements for motors in 430.14.

3. Section 430.14(B) requires motors (in this case, generators) having commutators or collector rings to be located so that sparks cannot reach adjacent combustible material. Answer B is true. Section 430.14 does not prohibit the installation on wooden floors. Answer A is not true.

4. Section 430.14(A) requires adequate ventilation and location to readily permit maintenance, such as lubrication of bearings. Answers C and D are true.

5. The correct answer is A.

34

A generator set has been determined to be vital to the operation of an electrical system. Instead of interrupting the generator circuit, the overload sensing device may:

A. Be omitted
B. Be interlocked
C. Be connected to an annunciator
D. Interrupt the generator field circuit

ANSWER ______________

PROCEDURE TO ESTABLISH ANSWER

The question is about means to interrupt current to the generator circuit.

1. In Index, find "Generators," under which, find "Overcurrent protection, 445.12."

2. In 445.12, the exception to (A) through (E) provides that where all the conditions have been met, the overload sensing device may to be connected to an annunciator or alarm.

3. The phrase "which is vital to operation of an electrical system" has an air of urgency or exception to the rule.

4. The correct answer is C.

35 Conductors between generator terminals and the first overcurrent device that carry the total current output of a generator shall have an ampacity of at least what percentage of the current rating of the generator?

A. 90
B. 100
C. 115
D. 125

ANSWER ______________

PROCEDURE TO ESTABLISH ANSWER

The question is about the sizing of generator conductors.

1. In Index, find "Generators," under which, find "Conductor, ampacity of, 445.13."
2. Section 445.13(A) requires the ampacity of the phase conductors from the generator terminals to the first overcurrent device be not less than 115 percent of the nameplate current rating of the generator, unless allowed by the exception, of which there is no mention here.
3. The correct answer is C.

36 A 200-kW, 480Y/277-volt, 3-phase, 4-wire generator set is part of a listed assembly, which includes a cabinet with the controls and main circuit breaker. Listed load shedding equipment is also part of the installation. Which one of the following statements is NOT true (including exceptions)?

A. The phase conductors from the generator terminals to the circuit breaker must be sized 115 percent of the nameplate current rating.
B. Where necessary for the safety of attendants, suitable insulating mats are required.
C. A minimum of a 1 m (3½ ft) working space is required between the control panel and a concrete wall opposite the panel.
D. A bushing is required where wires pass through an opening in the enclosure.

ANSWER ______________

PROCEDURE TO ESTABLISH ANSWER

The question is about generator conductor ampacity and installation procedures.

1. In Index, find "Generators," under which, find "Conductor, ampacity of, 445.13."
2. Section 445.13(A), Exception, permits the ampacity to be less than the required 115 percent but not less than 100 percent where the design and operation of the generator prevents overloading. Listed load shedding equipment is part of the installation and will prevent overloading. Answer A is not true.
3. In Index, under "Generators," also find "Guards for attendants, 445.15," which references Section 430.233 where necessary for the safety of attendants.
4. Section 430.233 requires insulating mats under certain conditions for safety of attendants. Answer B is true.

5. In Index, find "Working space," under which, find "About electrical equipment, 110.26, etc." Section 110.26(A)(1) refers us to Table 110.26(A)(1), which requires a 1.0 m (3½ ft) working space between equipment operating between 151–600 volts and a grounded surface such as a concrete wall. Answer C is true.

6. In Index, under "Generators" also find "Bushings, 445.16," which requires a bushing where wires pass through an opening in an enclosure. Answer D is true.

7. The correct answer is A.

37

A generator supplying an emergency system in a 30 m (100 ft) high apartment building may be installed in any of the following locations EXCEPT:

A. A room fully protected by an automatic water sprinkler system
B. A room fully protected by an automatic carbon dioxide fire protection system
C. A room with a 2-hour fire rating
D. A nonfire-rated electrical equipment room

ANSWER ______________

PROCEDURE TO ESTABLISH ANSWER

The question is about the location of emergency generators.

1. In Index, find "Generators," under which, find "Emergency systems, 700.12(D) ."

2. Section 700.12(D) does not provide any information for this question so scan Section 700.12.

3. Section 700.12(B), second paragraph requires that emergency sources of power including generators located in buildings (including residential) above 23 m (75 ft) in height be installed in spaces fully protected by approved automatic fire suppression systems (sprinklers, carbon dioxide, etc.) or in spaces with a 2-hour rating. (General Note: In the first paragraph, requirements for the equipment to be located so as to minimize the hazards from flooding, fires, icing, and vandalism are also mentioned).

4. The correct answer is D.

38

What is the minimum size of Type THW copper conductors required to supply a load consisting of a 5-horsepower motor and a 1000-watt resistive appliance on one branch circuit? The circuit operates at 240 volts, single-phase.

A. 10 AWG
B. 8 AWG
C. 6 AWG
D. 1 AWG

ANSWER ______________

PROCEDURE TO ESTABLISH ANSWER

The question is about sizing of branch circuits for a motor and combination of loads.

1. In Index, find "Motors," under which, find "Branch circuits," under which, find "Combination loads,

430.54, 430.63, 430.131." Also under "Motors," find "Conductors," under which, find "Combination loads, 430.25."

2. Section 430.25 requires conductors supplying a combination load, not factory wired, to have an ampacity determined in accordance with 430.24.

3. Section 430.24 requires the conductors to have an ampacity equal to the sum of motor full-load currents plus 125 percent of highest rated motor plus ampere rating of other loads determined by 430.6(A), plus the ampacity required for the other loads.

4. Section 430.6(A) requires the full load current (FLC) rating of a motor to be taken from the Tables 430.247 through 430.250.

5. Table 430.248 rates a 5-horsepower single-phase 230-volt motor full-load current at 28 amperes.

6. Where the type of resistive load is not otherwise specified, 220.14(A) requires the rating to be that of the nameplate of the appliance served, which is 1000 VA.

   I = 1000 VA ÷ 240 V

   I = 4.2 A.

7. (28 A x 125% = 35 A) + 4.2 A = 39.2 amperes total load.

8. Table 310.16 and 110.14(C)(1)(a) require a minimum of an 8 AWG Type THW copper conductor.

9. The correct answer is B.

39

Which one of the following statements about motor control circuits is NOT true?

A. If one side of a motor control circuit is grounded, the control circuit must be arranged so that an accidental ground in the remote control devices will not start the motor.
B. The motor control circuit conductors are always considered as protected by the motor branch-circuit short-circuit and ground-fault protective device.
C. If one side of a motor control circuit is grounded, the control circuit must be arranged so that an accidental ground in the remote control devices will not bypass automatic safety shutdown devices.
D. Where damage to a motor control circuit would constitute a hazard, all conductors of such circuit that are outside the control device shall be suitably protected from physical damage.

ANSWER ______________

PROCEDURE TO ESTABLISH ANSWER

The question is about motor control circuits.

1. In Index, find "Motors," under which, find "Control circuits, 430– VI." You can also go to Figure 430.1 and find "Motor Control Circuits 430.71 through 430.75.

2. Scan Part VI and find "430.73, Protection of Conductor from Physical Damage." The conductors are required to be installed in a raceway or otherwise protected from physical damage. Answer D is true.

3. Continue on in Part VI, and find 430.74, "Electrical Arrangement of Control Circuits." Section 430.74 requires when one side of the motor control circuit is grounded, the motor control circuit shall be so ar-

ranged that an accidental ground in the remote-control devices will not start the motor or bypass automatic safety shutdown devices. This section also requires protection from physical damage. Answers A and C are true.

4. Overcurrent protection for motor control circuits is covered by 430.72(A)(2) and may require protection other than provided by the branch-circuit device, particularly if they extend beyond the motor control equipment enclosure. From the above two responses to the question, we know that Answer B is not true.

5. The correct answer is B.

40

The disconnecting means for a torque motor shall have an ampere rating of at least what percentage of the motor nameplate?

A. 115
B. 125
C. 140
D. 150

ANSWER ______________

PROCEDURE TO ESTABLISH ANSWER

The question is about motor disconnecting means.

1. In Index, find "Motors," under which, find "Disconnecting means, 430.75, 430.95, 430– IX, etc." or you can go to Figure 430.1 and find "Disconnecting means, 430.101 through 430.113."

2. Scan Article 430 Part IX, and find "430.110, Ampere Rating and Interrupting Capacity."

3. Section 430.110(B) covering torque motors requires the disconnecting means for a torque motor to have an ampere rating of at least 115 percent of the motor nameplate current.

4. The correct answer is A.

41

At which of the following ratings is a motor disconnecting means for an ac motor permitted to be a general-use switch or isolating switch marked "Do not operate under load"?

A. Over 50 horsepower
B. Less than 75 horsepower
C. Over 100 horsepower
D. Less than 200 horsepower

ANSWER ______________

PROCEDURE TO ESTABLISH ANSWER

The question is about motor disconnecting means.

1. In Index, find "Motors, Art. 430," under which, find "Disconnecting means, 430.75, 430.95, 430– IX, 430.128, 430.227." or you can go to Figure 430.1 and find "Disconnecting means, 430.101 through 430.113."

2. Scan Article 430, Part IX, and find "430.109, Type." This section generally requires a horsepower-rated switch.

3. 430.109(C) exempts motors below 300 volts and 2 horsepower.

4. 430.109(E) permits the disconnecting means for ac motors over 100 horsepower to be a general use or isolating switch and plainly marked "Do not operate under load."

5. The correct answer is C.

42 Where motor branch-circuit short-circuit and ground-fault protection is provided by time-delay (dual-element) fuses and the rating is not sufficient to start the motor, it shall be permitted to increase the rating to what percentage of the full-load current rating of the motor?

A. 150%
B. 175%
C. 225%
D. 400%

ANSWER ______________

PROCEDURE TO ESTABLISH ANSWER

The question is about motor branch-circuit, short-circuit and ground-fault protection.

1. In Index, find "Motors," under which, find "Short circuit protection, 430– IV, 430– V, 430.130." You can also go to Figure 430.1 and find "Motor Branch-Circuit Short-Circuit and Ground-Fault Protection" 430.51 through 430.58."

2. Scan Article 430, Part IV, and find "430.52, Rating or Setting for Individual Motor Circuit."

3. Section 430.52(C)(1), Exception No. 2(b), allows a time-delay fuse to be increased to no more than 225 percent of the full-load current of a motor where the rating specified in Table 430.52 is not sufficient to allow the motor to start.

4. The correct answer is C.

43 The full-load current (FLC) rating of a 5-horsepower, single-phase, 208-volt motor is:

A. 15.2 amperes
B. 16.7 amperes
C. 28.0 amperes
D. 30.8 amperes

ANSWER ______________

PROCEDURE TO ESTABLISH ANSWER

The question is about full-load current ratings of motors.

1. In Index, find "Motors," under which, find "Current, full load. *see* Full-load current motors."

2. Referring back to Index, find "Full-load current motors," under which, find "Alternating current," under which, find "Single-phase, Table 430.248."

3. Table 430.248 specifies a full-load current of 30.8 amperes.

4. The correct answer is D.

   Note: Section 430.6(A)(1) requires this value (full-load current) to be used to size most of the components of the motor circuit. However, motor overload protection is specific to the motor used so it must be based on the actual motor nameplate ampere rating. The full-load current (FLC) values from the motor tables will normally be higher than on the nameplate.

44 Two or more motors, one horsepower or less, may be connected to a single 20-ampere, 120-volt branch circuit if all of the following conditions are met EXCEPT:

A. The motor is protected by time-delay fuses.
B. The full-load rating of each motor does not exceed 6 amperes.
C. The rating of the branch-circuit, short-circuit and ground-fault protective device marked on any of the controllers is not exceeded.
D. Individual overload protection conforming to 430.32 is provided.

ANSWER ______________

PROCEDURE TO ESTABLISH ANSWER

The question is about branch-circuit protection for more than one motor on a single circuit.

1. In Index, find "Motors," under which, find "Grouped, 430.24, 430.42, 430.53, 430.87, 430.112 Ex."

2. Section 430.53 is applicable since the relates to motors not over 1 horsepower where several motors are connected to one branch circuit.

3. Section 430.53(A) permits this installation where several motors are connected to a 120-volt, 20-ampere circuit and several conditions are met. Answers B, C, and D are true.

4. There is no requirement for the type of protection required. Fuse or circuit breaker protection would be allowed.

5. The correct answer is A.

45 A motor control enclosure that is located outdoors where rain, snow, and sleet are all likely, and must also meet the requirements to protect persons from incidental contact with the enclosed parts is required to have what minimum rating?

A. 1
B. 4X
C. 3S
D. 6P

ANSWER ______________

PROCEDURE TO ESTABLISH ANSWER

This is about rating of motor controller enclosures and their proper selection.

1. In Index, find "Enclosures," under which, find "Types, 110.28."

2. Section 110.28 refers us to Table 110.28, "Enclosure Types."

3. Table 110.28 indicates that a Type 3S is the minimum rating permitted by *Code* for the outdoor environment specified in the question.

4. The correct answer is C.

46 A central heating and air conditioning unit has the following information on the nameplate: "Minimum Circuit Ampacity — 28 Amps; Maximum Fuse Size — 45 Amps." What is the minimum size Type THW conductor that may be used to supply this unit?

A. 12 AWG
B. 10 AWG
C. 8 AWG
D. 6 AWG

ANSWER _______________

PROCEDURE TO ESTABLISH ANSWER

The question is about air conditioner branch-circuit sizing.

1. In the Index, either under "Air conditioning and refrigerating equipment, Branch circuits" and "Conductors" and find a reference to 440– IV. Realizing that quite often this type of unit typically contains a single motor compressor, you could find also find "Single motor compressor, 430.32" under "Branch circuits " in "Air conditioning and refrigerating equipment."

2. Scan Article 440, Part IV, and find "440.32," which specifies the procedure that the manufacturer uses to come up with the minimum circuit ampere provided in the question. Section 440.35 for multimotor and combination equipment refers to 440.4(B). 28 amps x 125% = 35 amps.

3. Section 440.4(B) requires the manufacturer to label the equipment with the information provided in the question.

4. Referring to Table 310.16 , the ampacity of 10 AWG Type THW is 35 amperes, and 110.14(C)(1)(a)(4) permits 75°C ampacity column in Table 310.16 which is the minimum size for the required 28-ampere circuit ampacity. This ampacity is permitted by 240.4(G) and Table 240.4(G).

5. The correct answer is B.

   Note: The maximum fuse size is a separate item. The conductors are sized based on the "minimum circuit ampacity," not on the (larger) "maximum fuse size."

47

Branch-circuit conductors supplying a single motor shall have a current-carrying capacity of not less than what percentage of full-load current of motor?

A. 115%
B. 125%
C. 175%
D. 250%

ANSWER _______________

PROCEDURE TO ESTABLISH ANSWER

The question is about sizing of branch-circuit conductors to a single motor.

1. In Index, find "Motors," under which, find "Conductors" under which find "Single motor, 430.22."
2. Section 430.22 requires branch-circuit conductors supplying a single motor to have an ampacity not less than 125 percent of the motor full-load current (FLC) rating.
3. The correct answer is B.

48

Which one of the following is the minimum permitted size of a Type THW copper branch-circuit conductor supplying a motor with a full-load current of 30 amperes?

A. 14 AWG
B. 12 AWG
C. 10 AWG
D. 8 AWG

ANSWER _______________

PROCEDURE TO ESTABLISH ANSWER

The question is about sizing motor branch-circuit conductors.

1. In Index, find "Motors," under which, find "Conductors" under which find "Single motor, 430.22."
2. Section 430.22 requires branch-circuit conductors supplying a single motor to have an ampacity not less than 125 percent of the motor full-load current rating.
3. 30 amps x 125% = 37.5 A, which is the minimum ampacity of motor branch-circuit conductors.
4. Referring to ampacity Table 310.16 , 8 AWG Type THW copper is rated 50 amperes. Keep 240.4(G) in mind. A 10 AWG Type THW copper conductor is rated at only 35 amperes
5. The correct answer is D.

49

A continuous-duty motor, rated 3 horsepower without a thermal protector integral with the motor, having a service factor of 1.15, shall be protected against overload by a separate overload protective device selected to trip at not more than what percentage of the motor nameplate full-load ampere (FLA) rating?

A. 115
B. 125
C. 140
D. 170

ANSWER ______________

PROCEDURE TO ESTABLISH ANSWER

The question is about motor overload protection.

1. In Index, find "Motors," under which, find "Overload protection, 430– III." You can also go to Figure 430.1 and find "Motor and Branch-Circuit Overload Protection 430.31 through 430.44."
2. Scan Article 430, Part III, and find "430.32," under which, find "430.32(A), More than 1 horsepower."
3. Section 430.32(A)(1) states a motor having not less than a 1.15 service factor is to be protected at no more than 125 percent of the motor full-load current rating.
4. The correct answer is B.

50

What maximum percentage shall be used for overload protection to protect motors of one horsepower, marked not over 40°C temperature rise and that start automatically?

A. 100
B. 115
C. 125
D. 140

ANSWER ______________

PROCEDURE TO ESTABLISH ANSWER

The question is about motor overload protection.

1. In Index, find "Motor," under which, find "Overload protection, 430– III." You can also go to Figure 430.1 and find "Motor Control Circuits 430.31 through 430.44."
2. Scan Article 430, Part III, and find "430.32(B)(1)," which permits a 1 horsepower or less motor with a marked temperature rise not over 40°C to be protected in accordance with 430.32(A)(1). Section 430.32(A)(1) permits a separate overload device under the conditions specified in the to be sized at 125 percent of the motor name plate full-load current rating.
3. The correct answer is C.
   Note: Do not confuse *40ºC temperature rise* with *40ºC ambient.* Many motors will be marked for use in a 40ºC ambient temperature rise. Many motors are generally rated for a 40ºC temperature rise over ambient.

51

What is the maximum allowable size of a motor branch-circuit protective device that provides overcurrent protection for a 14 AWG copper motor control-circuit conductor that extends from a motor controller to a start-stop push-button station located remotely from the controller?

A. 15 amperes
B. 45 amperes
C. 60 amperes
D. 100 amperes

ANSWER ______________

PROCEDURE TO ESTABLISH ANSWER

The question is about overcurrent protection for motor control-circuit conductors.

1. In Index, find "Motors," under which, find "Control circuits, 430– VI." You can also go to Figure 430.1 and find "Motor Control Circuits 430.71 through 430.75.
2. Scan Article 430, Part VI, and find "430.72, Overcurrent Protection," and "Table 430.72(B), Maximum Rating of Overcurrent Protective Device-Amperes."
3. 430.72(B)(2) states that conductors extending beyond the motor control equipment enclosure are permitted to be protected by the motor branch-circuit short-circuit and ground-fault protective devices where the rating is not over the value specified in Column C of Table 430.72(B).
4. Column C of Table 430.72(B) indicates a maximum 45-ampere rating for 14 AWG copper.
5. The correct answer is B.

52

A 3-phase, 10-horsepower motor is operated from a 240-volt, 3-wire, 3-phase delta source, which has one phase grounded. The minimum number of fuses required for overload protection shall be _____?

A. 0
B. 1
C. 2
D. 3

ANSWER ______________

PROCEDURE TO ESTABLISH ANSWER

The question is about motor overload protection.

1. In Index, find "Motors," under which, find "Overload protection, 430– III." You can also go to Figure 430.1 and find "Motor and Branch-Circuit Overload Protection 430.31 through 430.44."
2. Scan Article 430, Part III, and find "430.36," which provides that where fuses are used for motor overload protection, a fuse shall be inserted in each ungrounded conductor. The second part of the sentence also requires three fuses where a 3-wire, 3-phase ac system has one conductor grounded.
3. The correct answer is D.

53 All of the following statements about a motor disconnecting means and controller are true EXCEPT:

A. A control circuit push-button station located within sight from the controller may be used as the motor disconnecting means.
B . The disconnecting means shall be located in sight from the controller.
C. A disconnecting means capable of being locked in the open position may be located out-of-sight from the motor location.
D. A combination lockable disconnecting means and controller located in a motor control center may be located out-of-sight from the motor location.

ANSWER ______________

PROCEDURE TO ESTABLISH ANSWER

The question is about the location of motor disconnecting means.

1. In Index, find "Motors," under which, find "Disconnecting means, 430.75, 430.95, 430– IX, etc."

2. Scan Article 430, Part IX, and find "430.102, Location."

3. Section 430.102(A) requires the disconnecting means to be in sight from the controller location. Answer B is true.

4. Section 430.102(B), Exception, permits the disconnecting means to be out-of-sight from the motor if it is capable of being locked in the open position and either of the stated conditions are met. Answers C and D are true.

5. The correct answer is A.
Note: Section 430.103 requires the disconnecting means to open all ungrounded supply conductors. A control circuit switch does not open the ungrounded conductors but only power for the control circuit , so it does not qualify as a motor disconnecting means. The disconnecting means is not permitted to be closed automatically, such as with a time switch.

54 The minimum number of overload units providing overload protection for a 3-phase ac motor connected to a 208Y/120-volt, 3-phase, 4-wire system is:

A. 1
B. 2
C. 3
D. 4

ANSWER ______________

PROCEDURE TO ESTABLISH ANSWER

The is about requirements for motor overload units.

1. In Index, find "Motors," under which, find "Three overload units, Table 430.37."

2. For an ac motor, Table 430.37 requires three devices, one in each phase.

3. The correct answer is C.

55

Where the maximum permitted motor branch circuit short-circuit and ground-fault protective device ratings are not sufficient for the starting of the motor, which one of the following statements is true?

A. The next higher rated overcurrent device cannot be used.
B. The rating may be increased to 400 percent for time-delay fuses.
C. The rating may be increased as necessary to permit starting.
D. The manufacturer's recommendations or markings shall not be exceeded, even if higher values are shown in the *Code*.

ANSWER _______________

PROCEDURE TO ESTABLISH ANSWER

The question is about the rating of the motor branch-circuit and short-circuit devices.

1. In Index, find "Motors," under which, find "Overcurrent protection," under which, find "Branch circuits, 430– IV." You can also go to Figure 430.1 and find "Motor Branch-Circuit Short-Circuit and Ground-Fault Protection, 430.51 through 430.58."

2. Since the question is concerned with changing the rating or setting of the motor branch-circuit protective device, scan Article 430 Part IV and find "430.52, Rating or Setting for Individual Motor Circuit."

3. Section 430.52(C)(2) requires that the manufacturer's marked rating not be exceeded. Answer D is true. Read through 430.52 and find that the other three answers are incorrect.

   Section 430.52(C)(1), Exception No. 1 permits the use of the next standard size when the values determined by Table 430.52 do not correspond with a standard rating. Answer A is not true. The provision provided in Exception No. 2 for a time-delay fuse is 225%, not 400%. Answer B is not true. There are limits to increasing the overcurrent protection. The device ratings may not be simply increased at will. Answer C is not true.

4. The correct answer is D.

56

Two motors each have a full-load current of 50 amperes and each motor branch-circuit is protected by a 150-ampere, non-time-delay fuse. The feeder overcurrent device shall not be larger than:

A. 62.5 amperes
B. 150 amperes
C. 200 amperes
D. 250 amperes

ANSWER _______________

PROCEDURE TO ESTABLISH ANSWER

The question is about motor feeder overcurrent protection.

1. In Index, find "Feeders," under which, find "Overcurrent protection, motor feeders, 430– V."

2. Scan Article 430, Part V, and find "430.62(A)," which limits the feeder protective device to the rating of the largest branch-circuit short-circuit and ground-fault protective device of the group (based on Table

430.52), plus the sum of the full-load currents (FLC) of the other motors of the group.

Where the same rating or setting of the branch-circuit short circuit and ground-fault protective device is used on two or more of the branch circuits supplied by the feeder, one of the protective devices shall be considered the largest for the above calculations.

3. 50A FLC x 300% = 150 A (motor 1)
   150 A + 50 A (motor 2) = 200 A

4. The correct answer is C.

57

Which one of the following statements is NOT true?

A. An attachment plug and receptacle is permitted to be used as a disconnecting means for a portable motor rated 1/3 horsepower or less.

B. For a stationary motor rated 1/8 horsepower or less and constructed so that it can't be damaged by overload or failure to start, such as a clock that is normally left running, the branch-circuit protective device is permitted to be the controller.

C. For a portable motor rated ½ horsepower, the controller shall be permitted to be an attachment plug and receptacle.

D. For a cord- and plug-connected motor, a horsepower-rated attachment plug and receptacle having ratings no less than the motor ratings shall be permitted to serve as the disconnecting means.

ANSWER ______________

PROCEDURE TO ESTABLISH ANSWER

The question is about an attachment plug and receptacle used as a disconnecting means and controller for portable motors, and motor controllers, in general, for smaller motors.

1. In Index, find "Motors," under which, find "Disconnecting means, 430.75, 430.95, 430– IX, etc."

2. In Index, find "Motors," under which, find " *see* Controllers, motor." Under "Controllers" in the Index, find "Motor, see motor controllers" so it sends you in a circle without giving you the location. You can go to Figure 430.1 at the beginning of Article 430 and find "Motor Controllers 430.81 through 430.90."

3. Scan Article 430, Part VII, and find "430.81(B)," which provides that for portable motors rated 1/3 horsepower or less, an attachment plug and receptacle or cord connector is permitted to serve as a controller. Answer A is true, and Answer C is not true.

4. Section 430.81(A) permits stationary motors rated ⅛ horsepower or less to be protected by the branch-circuit protective device. Answer B is true. For example, this could be a small clock motor or group of clock motors in a school.

5. Scan Article 430, Part IX, and find "430.109," or use Figure 430.1 and find "Disconnecting Means 430.101 through 430.113" under which 430.109(F) permits a horsepower-rated attachment plug and receptacle. Answer D is true.

6. The correct answer is C.

58 All of the following statements about motor controllers are true EXCEPT:

A. A controller not serving as a disconnecting means is required to open all conductors to the motor.
B. A motor controller shall be capable of interrupting the locked-rotor current of the motor.
C. Each controller must be capable of starting and stopping the motor that it controls.
D. Where the controller also serves as a disconnecting means, it shall open all ungrounded conductors to the motor.

ANSWER _______________

PROCEDURE TO ESTABLISH ANSWER

The question is about the performance of motor controllers.

1. You can also go to Article 430 in the Table of Contents and find "Part VII, Motor Controllers, Page 70-323 " which will take you "Part VII Motor Controllers" and to 430.82 or use Figure 430.1 and find "Motor Controllers 430.81 through 430.90."

2. Scan Article 430, Part VII, and find "430.82," which requires the controller to be capable of starting and stopping the motor it controls and be capable of interrupting locked-rotor current of the motor. Answers B and C are true.

3. Scan Article 430, Part VII and find "430.84," which states the controller need not open all ungrounded conductors, except where it also serves as the disconnecting means. Answer D is true and Answer A is not true.

4. The correct answer is A.

59 The controller for a 25-horsepower motor shall have a horsepower rating not less than what percentage of the horsepower-rating of the motor?

A. 100
B. 115
C. 125
D. 150

ANSWER _______________

PROCEDURE TO ESTABLISH ANSWER

The question is about ratings of motor controllers.

1. In Index, find You can go Article 430 in the Table of Contents and find "Part VII, Motor Controllers, Page 70-312" which will take you "Part VII Motor Controllers" or use Figure 430.1 and find "Motor Controllers 430.81 through 430.90."

2. Scan Article 430, Part VII, and find "430.83(A)," which requires controllers, other than inverse-time circuit breakers and molded-case switches, to have a horsepower-rating at the application voltage not lower than the horsepower-rating of the motor.

3. The correct answer is A.

60 A 480Y/277-volt motor control center (MCC) is located opposite a 208Y/120-volt panelboard. What is the minimum working space between the MCC and the panelboard?

A. 900 mm (3 ft)
B. 1.0 m (3½ ft)
C. 1.2 m (4 ft)
D. 1.8 m (6 ft)

ANSWER ______________

PROCEDURE TO ESTABLISH ANSWER

The question is about working space in front of a motor control center.

1. In Index, find "Working space," under which, find "About electrical equipment, 110.26, 110.32, etc."
2. Section 110.26(A) requires working clearances to be not less than that in Table 110.26(A)(1). Since the MCC is opposite electrical equipment, Condition 3 applies.
3. With the MCC operating at 480 volts, the voltage to ground would be in the 151 – 600 range and the minimum clearance is 1.2 m (4 ft).
4. The correct answer is C.

61 The maximum distance permitted when the term "in sight from" as used in 430.102 is:

A. 3.0 m (10 ft)
B. 7.5 m (25 ft)
C. 15 m (50 ft)
D. 23 m (75 ft)

ANSWER ______________

PROCEDURE TO ESTABLISH ANSWER

The question is about the maximum distance permitted by the definition of *in sight from*.

1. In Index, find "In sight from," under which, find "Definition, Art. 100– Part I."
2. In Article 100, Definitions, find *In sight from* which states that the specified equipment is to be visible and not more than 15 m (50 ft) distance from the other piece of equipment — such as between a motor and its disconnect.
3. The correct answer is C.

62

A disconnect switch for a stationary ac motor need not be horsepower-rated if the motor is larger than what horsepower?

A. 25
B. 50
C. 75
D. 100

ANSWER ______________

PROCEDURE TO ESTABLISH ANSWER

The question is about the rating of a motor disconnect switch.

1. In Index find "Motors," under which, find "Disconnecting means, 430.75, 430.95, 430– IX, etc." or use Figure 430.1 and find "Disconnecting Means 430.101 through 430.113."

2. Scan Article 430, Part IX, and find "430.109(A)(1)," which requires the disconnecting means to be a motor-circuit switch rated in horsepower, etc.

3. However, 430.109(E) permits the disconnecting means for a stationary ac motor rated over 100 horsepower to be other than horsepower-rated, where it is a general use or isolating switch and plainly marked "Do not operate under load."

4. The correct answer is D.

63

The total load on an individual branch circuit for a cord- and attachment-plug-connected room air conditioner cannot exceed what percentage of the branch-circuit load?

A. 50
B. 80
C. 115
D. 125

ANSWER ______________

PROCEDURE TO ESTABLISH ANSWER

The question is about a branch circuit for room air-conditioning units.

1. In Index, find "Air conditioners, rooms," under which, find "Branch-circuit requirements, 440.62."

2. Section 440.62(B) states where no other loads are also supplied, the A/C room unit load shall not exceed 80 percent of the rating of the branch circuit.

3. The correct answer is B.

64 The service overcurrent protection for a fire pump motor shall comply with which one of the following?

A. Be selected to protect the motor from dangerous overheating
B. Be selected or set to carry locked-rotor current indefinitely
C. Be selected according to the requirements for motor overload protection
D. Be selected according to the requirements for motor branch-circuit protection

ANSWER _______________

PROCEDURE TO ESTABLISH ANSWER

The question is about overcurrent protection for fire pump motors.

1. In Index, find "Fire pumps," under which, find "Motor, overcurrent protection, 430.31."

2. Section 430.31 provides that *overload protection* is not required for fire pump circuits, but our concern here is the broader of overcurrent protection, including service overcurrent protection. The Informational Note sends readers of the *Code* to 695.7 for protection of fire pump supply conductors; however, ours is related to overcurrent device selection.

3. In Index, find "Fire pumps," under which, find "Service equipment overcurrent protection, 230.90(A), Ex. 4." Exception No. 4 provides that fire pump service overload protection complies with 695.4(B)(2)(a), which states overcurrent devices are required to be selected to carry locked-rotor currents indefinitely; and if the locked-rotor current does not match the size of a standard overcurrent protective device, the next higher size shall be used in accordance with 240.6.

4. The correct answer is B.

65 Which one of the following is required to be based on the actual motor nameplate current rating?

A. Motor overload protection
B. Motor branch-circuit short-circuit and ground-fault protection
C. Size of motor controller
D. Size of motor disconnecting means

ANSWER _______________

PROCEDURE TO ESTABLISH ANSWER

The question is about ratings of motor circuit equipment.

1. In Index, find "Motors, Art. 430," under which, find (for the lack of a directly related code section) "General, 430– Part I."

2. Scan Article 430, Part I, and find "430.6(A)(1), Table Values." The answers listed in B through D above are required to be sized based on the values given in the Article 430 Tables 430.247 through 430.250, as applicable.

3. Section 430.6(A)(2) requires the nameplate current rating be used only to size separate motor overload protection.

4. The correct answer is A.

66 Which of the following statements about the class of motor permitted in a Class I, Zone 2 hazardous (classified) area is true?

A. It must be listed for a Class I, Zone 2 area.
B. It must be listed for a Class I, Division 1 area.
C. It must be listed for a Class I, Division 2 area.
D. An open, squirrel-cage induction motor without brushes or a switching mechanism is permitted.

ANSWER ______________

PROCEDURE TO ESTABLISH ANSWER

The question is about the type of motors that are permitted in a Class I, Zone 2 hazardous (classified) area.

1. In Index, find "Motors, Art. 430," under which, find "Hazardous (classified) locations, 501.125, 502.125, 503.126, 505.16(E)(2), 505.22."

2. Since the location in the question is a Class I, Zone 2 area, recognize that additional research must be done since Article 501 applies to Class I, Division 1 and 2 locations (not Class I, Zone 2 locations), Article 502 applies to Class II locations and Article 503 applies to Class III locations.

3. In Index, find "Hazardous (classified) locations, Class I, Zone 0, 1 and 2 locations, Art. 505." Since there is no subheading for "motors," note the entry for "Equipment for use in, 505.20."

4. Scan Section 505.20 and find in 505.20(C) Exception No. 4 that in Zone 2 areas, the *Code* permits "the installation of open or nonexplosionproof or nonflameproof enclosed motors, such as squirrel-cage induction motors without brushes, switching mechanisms, or similar arc-producing devices that are not identified for use in a Class I, Zone 2 location...."

5. The correct answer is D.

67 What are the maximum locked rotor current amperes of a 40-horsepower, 3-phase code letter G, 460-volt (actual voltage for the circuit is 480-Volts) motor?

A. 166 amperes
B. 480 amperes
C. 303 amperes
D. 326 amperes

ANSWER ______________

PROCEDURE TO ESTABLISH ANSWER

The question is about finding the locked rotor current of a 3-phase motor. Remember, these current values are different from the full load amperes. Locked rotor currents are present in a motor circuit typically when the motor is starting, or when the motor rotor is actually locked up and the shaft will not move due to a mechanical problem or failure.

1. As you scan through "Motors" in the Index, you'll notice that "Locked Rotor Current" is not found. Next, look in the Index for the term "*Locked Rotor Motor Current*" in its own category and you'll find a reference to "Code Letters, 430.7(B), Table 430.7(B)."

2. In 430.7(B), you are referred to Table 430.7(B), which lists a range of values for code letter G motors. The maximum value is 6.29 KVA per horsepower.

3. The formula to find 3-phase locked rotor current is as follows:
   Locked Rotor Amps = Motor HP x Max. Code Letter value (KVA/HP) x 1000 / Volts x 1.73

4. Place the component numbers into the equation in the proper locations:
   LRA = 40 HP x 6.29 (KVA/HP) x 1000 / 480 volts x 1.73

5. Solve the equation:
   LRA = 40 HP x 6.29 (KVA/HP) x 1000 / 480 volts x 1.73
   LRA = 251,600 VA ÷ 831.36 V_{eff}
   LRA = 302.77 amps (rounded up to 303 amperes)

6. The correct answer is C.

68 Motors equipped with electrically powered condensation heaters shall be marked with which of the following:

A. The rated heater voltage
B. Number of phases
C. Rated power in watts
D. All of the above

ANSWER ______________

PROCEDURE TO ESTABLISH ANSWER

This deals with information that is required to be marked on the motor.

1. In Index, find "Motors," under which, find "Marking, 430.7"

2. Scan 430.7(A)(15). This shows that A, B, and C are true. Therefore, Answer D is true.

3. The correct answer is D.

69 Generally, what information is required to be marked on a motor controller, other than the manufacturer's name or identification, or other necessary data to properly indicate applications for which it is suitable?

A. Voltage
B. Current- or horsepower-rating
C. Short-circuit current rating
D. All of the above

ANSWER ______________

PROCEDURE TO ESTABLISH ANSWER

The deals with what is required to be marked on a motor controller.

1. Under "Controllers," you will find "Motors, See Motor Controllers" and you will not find it under "Motors" in the index .

2. Look in the "General Section, 430– Part I" of Article 430, Motors. Scan through Part I, General, and find "430.8, Marking on Controllers." This shows that A, B, and C are true. Therefore, Answer D is true.

3. The correct answer is D.

CHAPTER 5
CABINETS

1 A panelboard contains 120/240-volt, multiwire branch-circuits, protected with 30-ampere or less single-pole overcurrent devices. What is the maximum number of overcurrent devices permitted in any one cabinet or cutout box?

A. 18
B. 42
C. 30
D. The maximum number of overcurrent devices for which the panelboard was designed.

ANSWER ______________

PROCEDURE TO ESTABLISH ANSWER

The question is about the number of overcurrent devices in panelboards.

1. In Index, find "Panelboards." Unfortunately, in the Index, there is no good reference to find the maximum number of overcurrent devices permitted in a panelboard. If you concluded the question was relevant to overcurrent protection, you could find "Panelboards" in the Index under which find "Overcurrent protection, 408.36, 408.52, 408.54 etc." Other than as specified in 408.36 Exception No. 1 and by any manufacturer's requirements, there are no restrictions on the number of overcurrent devices in a cabinet.

   Section 408.54 states that panelboards are to be provided with physical means to prevent the installation of more overcurrent devices than the number for which the panelboard was designed, rated and listed.

2. The correct answer is D.

2 Without considering any exceptions, the minimum height permitted for working spaces for 480-volt service equipment, switchboards, panelboards, or motor control centers is what distance?

A. 1.5 m (5 ft)
B. 1.8 m (6 ft)
C. 2.0 m (6 ½ ft)
D. 2.1 m (7 ft)

ANSWER ______________

PROCEDURE TO ESTABLISH ANSWER

The question is about headroom about electrical equipment.

1. In Index, find "Working space," under which, find "About electrical equipment, 110.26, 110.32 through 110.34, etc."

2. Section 110.26(A)(3) requires a minimum of 2.0 m (6½ ft) the height of the equipment, whichever is greater.

3. The correct answer is C.

3 Raceways entering the bottom of a floor-standing switchboard, motor control center or similar enclosure shall not be more than what distance above the bottom of the enclosure?

A. 75 mm (3 in.)
B. 150 mm (6 in.)
C. 200 mm (8 in.)
D. 250 mm (10 in.)

ANSWER ______________

PROCEDURE TO ESTABLISH ANSWER

The question is about conductors entering enclosures such as switchboards.

1. In Index, find "Switchboards, Art. 408," under which, find "Clearances, 110.26, 408.5, 408.18"

2. In Section 408.5, Clearance for Conductors Entering Bus Enclosures," find that conduit and raceways, including end fittings, shall not rise more than 75 mm (3 in.) above the bottom of the enclosure.

3. The correct answer is A.

4 What is the minimum air space permitted between a metal cabinet and the supporting surface in a damp or wet location?

A. 3 mm (⅛ in.)
B. 6 mm (¼ in.)
C. 13 mm (½ in.)
D. 19 mm (¾ in.)

ANSWER ______________

PROCEDURE TO ESTABLISH ANSWER

The question is about cabinets and cutout boxes in damp or wet locations.

1. In Index, find "Cabinets, cutout boxes, and meter socket enclosures," under which, find "Damp and wet, 312.2."

2. Section 312.2 requires metal cabinet or cutout boxes located in damp or wet locations to be mounted so there is at least 6-mm (¼-in.) air space between the enclosure and the wall or supporting surface. The exception permits nonmetallic enclosures to be installed without the airspace on concrete, masonry, or tile surfaces, but not where rust of metal or rotting of wood may occur if no airspace exists between the enclosure and the mounting surface.

3. The correct answer is B.

5 Parallel 250-kcmil conductors (two per phase) enter a cabinet containing a panelboard, as indicated in the drawing. What is the minimum wire bending space required at the terminals?

A. 114 mm (4½ in.)
B. 152 mm (6 in.)
C. 203 mm (8 in.)
D. 254 mm (10 in.)

ANSWER ______________

PROCEDURE TO ESTABLISH ANSWER

The question is about deflection of conductors that do not enter or leave the enclosure wall opposite its terminals.

1. In Index, find "Cabinets, cutout boxes, and meter socket enclosures," under which, find "Wire-bending space at Terminals, 312.6(B)."

2. Section 312.6(B)(1) states Table 312.6(A) applies when the conductor does not enter or leave the enclosure through the wall opposite its terminal.

3. Refer to Table 312.6(A). Since the concern is parallel conductors, "Column 2, 2 wires per terminal," shall be used. Column 2 requires 152 mm (6 in.) of wire-bending space for parallel 250 kcmil conductors.

4. The correct answer is B.

6 Which wall of a cabinet does Table 312.6(B) address when concerned with conductors entering switches or panels for termination?

A. The top
B. The wall opposite its terminal
C. The side gutter space
D. The bottom

ANSWER ______________

PROCEDURE TO ESTABLISH ANSWER

The question is about deflection of conductors entering enclosures.

1. In Index, find "Conductors," under which, find "Cabinets and cutout boxes, 312.5 through 312.7." You can also find "Cabinets, cutout boxes, and meter socket enclosures," in the Index under which, find "Wire-bending space at terminals, 312.6(B)."

2. Section 312.6(B)(2) states Table 312.6(B) shall apply where conductors enter or leave the enclosure through the wall opposite its terminal.

3. The correct answer is B.

7 What is the minimum wire-bending space for two 4/0 AWG conductors installed in parallel that leave the enclosure through the wall opposite its terminal?

A. 165 mm (6½ in.)
B. 178 mm (7 in.)
C. 190 mm (7½ in.)
D. 203 mm (8 in.)

ANSWER ______________

PROCEDURE TO ESTABLISH ANSWER

The question is about multiple conductors per phase that leave an enclosure through the wall opposite the terminals.

1. In Index, find "Conductors," under which, find "Cabinets and cutout boxes, 312.5 through 312.7." You can also find "Cabinets, cutout boxes, and meter socket enclosures," in the Index under which, find "Wire-bending space at terminals, 312.6(B)."

2. Section 312.6(B)(2) states Table 312.6(B) applies where conductors enter or leave the enclosure through the wall opposite its terminal.

3. Refer to Table 312.6(B). Since the question refers to 4/0 AWG wire, in the column titled "Wire Size" refer to the "2 Wires per Terminal" column for a space of 190 mm (7½ in.).

4. The correct answer is C.

8 What is the minimum wire-bending space required for one 500-kcmil conductor at each lay in terminal in the panelboard shown in the illustration?

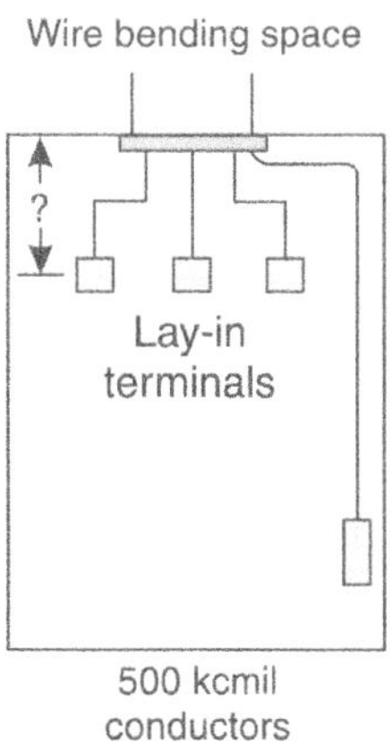

A. 152 mm (6 in.)
B. 279 mm (11 in.)
C. 330 mm (13 in.)
D. 356 mm (14 in.)

ANSWER ______________

PROCEDURE TO ESTABLISH ANSWER

The question is about conductors that enter or leave enclosure wall opposite its terminal.

1. In Index, find "Conductors," under which, find "Cabinets and cutout boxes, 312.5 through 312.7."

2. Section 312.6(B)(2) states that Table 312.6(B) shall apply where conductors enter or leave the enclosure through the wall opposite its terminal.

3. Refer to Table 312.6(B). Since the question refers to 500-kcmil wire, in column titled "Wire Size" refer to "1 per Terminal" for a space of 356 mm (14 in.).

4. Note the superscript "e" beside the dimensions, which refers to the notes at the bottom of the table.

5. Since the terminals are the lay-in type, Note 2 applies. Referring to the note below Table 312.6(B), you will find that the bending space may be reduced by 76.2 mm (3 in.) for those sizes indicated by the superscript "e."

   356 mm (14 in.) – 76.2 mm (3 in.) = 279 mm (11 in.)

6. The correct answer is B.

9 Where using a multi-section switchboard for service equipment, which one of the following statements is true?

A. Each section is required to have a main bonding jumper.
B. Each section must be bonded together with a conductor sized in accordance with Table 250.102(C)(1).
C. Each section must be connected together with an equipment-bonding jumper or supply-side bonding jumper, sized according to Table 250.122 or Table 250.102(C)(1) as appropriate.
D. Each section must have the grounded service conductor bonded to it.

ANSWER ______________

PROCEDURE TO ESTABLISH ANSWER

The question is about bonding sections of switchboards used for service equipment.

1. In Index, find "Switchboards, Art. 408."

2. Since a specific section is not given, scan sections of Article 408.

3. Section 408.3(C) requires that each switchboard shall be provided with a main bonding jumper in one section sized per 250.28(D). Answer A is not true as it references a main bonding jumper in each section.

   Section 403.3(C) also specifies that each section shall be bonded together with an equipment-bonding jumper or a supply-side bonding jumper, sized in accordance with Table 250.122 or Table 250.102(C)(1)' as appropriate. Answer B is not true. Answer C is true.

   The main bonding jumper is required for connection of the grounded conductor to the switchboard on its supply side. Therefore, the requirement is only for one location. Answer D is not true.

4. The correct answer is C.

10

All of the following statements about a metal metering cabinet mounted on a wall on the outside of a building exposed to the weather are true EXCEPT:

A. No airspace is not required between the cabinet and a concrete wall.
B. The cabinet must be installed to prevent water from entering or accumulating in the cabinet.
C. The cabinet must be of weatherproof construction.
D. The cabinet must be mounted so there is at least 6 mm (¼ in.) airspace between the cabinet and wall.

ANSWER _______________

PROCEDURE TO ESTABLISH ANSWER

The question is about mounting a metering cabinet on the outside of a building. Note that equipment exposed to the weather is considered to be in a wet location based on the Article 100 definition of *wet location.*

1. In Index, find "Cabinets, cutout boxes, and meter socket enclosures" under which find "Damp, and wet locations, 312.2."

2. Section 312.2 requires the enclosure to be placed to prevent water from entering and accumulating in the enclosure. Answer B is true. This section also requires at least a 6-mm (¼ in.) airspace between the wall and the enclosure. Answer D is true.

   As noted in Answer D, metallic metering enclosures require an air space, so Answer A is not true. The only exception is for nonmetallic enclosures.

   Since the location is considered to be a wet location, the cabinet is required to be weatherproof. Answer C is true. Answer A is not true, because this is a metal and not a nonmetallic cabinet.

3. The correct answer is A.

11

Which one of the following statements concerning cabinets is true?

A. Cabinets that contain devices or apparatus connected to more than eight conductors must have back or side wiring space.
B. Several NM cables are not permitted to enter cabinets through an open-ended single conduit.
C. Conductors shall not be deflected within a cabinet or cutout box unless a gutter having a width in accordance with Table 312.6(B) is provided.
D. Taps may fill the cross-sectional area of a wiring space no more than 60%.

ANSWER _______________

PROCEDURE TO ESTABLISH ANSWER

The question is about the construction of cabinets, cutout boxes and deflection of conductors.

1. In Index, find "Cabinets, cutout boxes, and meter socket enclosures," under which, find "Construction specifications, 312– II" and "Deflection of conductors, 312.6."

2. Scan Article 312, Part II, and find "312.11(C), Wiring Space," which requires cabinets that contain devices or apparatus connected to more than eight conductors to have back or side wiring space. Answer A is true.

3. Section 312.6(A) states conductors shall not be deflected, unless a gutter width in accordance with Table 312.6(A) is provided; conductors in parallel must be considered and require greater gutter width. Answer C states Table 312.6(B), which is not true.

4. Neither Part II nor 312.6 addresses answer B; however, Section 312.5, Cabinets, Cutout boxes and Meter Socket Enclosures, does address it. Section 312.5(C), Cables, permits cables with entirely nonmetallic sheaths to enter the top of a surface-mounted enclosure through a nonflexible raceway provided all of the conditions of the exception are complied with. Answer B is not true.

5. Section 312.8(A)(2) provides that taps may fill the cross-sectional area of a wiring space no more than 75 percent, per 312.8. Answer D is not true.

6. The correct answer is A.

12

Which one of the following statements concerning a new switchboard not containing metering equipment and supplied from a 3-phase, 4-wire, 120/240-volt delta ("high leg") system is true?

A. The high leg must be the center bus (B phase) of the three-phase busses.
B. Conductors are not permitted to pass horizontally through vertical sections of switchboards.
C. Main bonding jumpers are not permitted in switchboards.
D. The high leg bus must be identified by a red color.

ANSWER ______________

PROCEDURE TO ESTABLISH ANSWER

The question is about conductors in switchboards supplied from a delta-connected, 3-phase, 4-wire system.

1. In Index, find "Switchboards," under which, find "Support, busbars, conductors, 408.3."

2. Under 408.3, find "408.3(E), Bus Arrangement." Section 408.3(E)(1), AC Phase Arrangement, specifies the "B phase" must be that phase having the higher voltage to ground on a 3-phase, 4-wire delta-connected system.

3. Since the phase arrangement must be A, B, C, etc., the center bus must be the high leg. The exception permitting a different phase configuration applies only where a meter is in the same section. Answer A is true.

4. Note the Informational Note, which refers to 110.15 for requirements on marking the high leg of a 4-wire, delta-connected system. Answer D is not true.

5. Section 110.15 states that the higher phase voltage to ground shall be durably and permanently marked by an outer finish that is orange in color or by other effective means.

6. A review of Sections 408.3(A)(2) Exception and 408.3(C) confirms that answers B and C are not true. Conductors may pass through switchboards where barriers are provided to isolate persons from live busbars. A main bonding jumper is required (in a service switchboard, for instance.)

7. The correct answer is A.

13

Where insulated circuit conductors enter a cabinet through a raceway, the conductors shall be protected by an identified insulating bushing or other similar device where the conductors are:

A. 8 AWG or larger
B. 6 AWG or larger
C. 4 AWG or larger
D. 2 AWG or larger

ANSWER ______________

PROCEDURE TO ESTABLISH ANSWER

The question is about protection of conductors entering enclosures.

1. In Index, find "Conductors," under "Insulated", find "At bushings, 4 AWG and larger, 300.4(G), 312.6(C)."

2. Section 300.4(G) requires insulated circuit conductors of 4 AWG or larger entering a cabinet, box enclosure, or raceway to be protected by an identified fitting providing a smoothly rounded insulated surface, etc.

3. The correct answer is C.

14

Which one of the following statements about a conduit bushing constructed wholly of insulating material and used to protect conductors entering a cabinet is true?

A. The bushing alone shall not be used to secure the raceway.
B. The conductors must be separated where they pass through the bushing.
C. The bushing may be used to secure raceways.
D. The bushing is acceptable if a single locknut is installed on the outside of the cabinet.

ANSWER ______________

PROCEDURE TO ESTABLISH ANSWER

The question is about insulated conduit bushings.

1. In Index, find "Bushings," under which, find "Insulated, 300.4(G), 312.6(C)."

2. Section 300.4(G) states that conduit bushings constructed wholly of insulating material shall not be used to secure a raceway.

3. There is no recognition that statements B, C, or D are true as conduit bushings constructed wholly of insulating material shall not be used to secure a raceway.

4. The correct answer is A.

15

Which one of the following items is permitted in the space directly above an indoor panelboard?

A. A nonmetallic pipe for a pneumatic air system
B. A transformer providing power to the panelboard
C. A nonmetallic waste disposal drainpipe
D. A sheet-metal duct providing ventilation to the area

ANSWER ______________

PROCEDURE TO ESTABLISH ANSWER

The question is about the installation of a panelboard.

1. In Index, find "Panelboards," under which, find "Installation, 110.26(E)."

2. Section 110.26(E)(1) requires the space directly above the panelboard to a height of 1.8 m (6 ft) or the structural ceiling, whichever is lower, to be dedicated to the electrical installation. No piping, ducts, or

equipment foreign to the electrical installation is permitted in this space. The nonmetallic pipes and sheet-metal duct would be prohibited. Answers A, C and D are not true. The transformer is considered part of the electrical installation and is permitted in the space above the panelboard.

3. The correct answer is B.

   Note: A transformer located above or below the panelboard in the working space would not be permitted, if it projected out more than 150 mm (6 in.) into the workspace in front of the panel enclosure as specified in 110.26(A)(3). Make sure to read and understand the requirements for working spaces and dedicated spaces as it pertains to these transformer installations.

16

An enclosure designed either for surface or flush mounting and provided with a frame, mat, or trim in which a swinging door or doors are or may be hung is called a:

A. Junction box
B. Cabinet
C. Cutout box
D. Panelboard

ANSWER ______________

PROCEDURE TO ESTABLISH ANSWER

The question is about enclosures such as cabinets and cutout boxes.

1. In Index, find "Enclosures," under which, find "Cabinets and cutout boxes, 312.7 through 312.9, 312.11(D)." You can also find "Cabinets, cutout boxes, and meter socket enclosures," in the Index under which, find "Definitions, 100-I."

2. Definitions specific to an article may be located within that article in Section ".2" and scanning Article 312, there are no definitions of a cabinet or cutout box. Therefore, scan Article 100 for the list of above terms.

3. The definition of cabinet is found to comply with the question.

4. The correct answer is B.

17

Which one of the following statements about a cabinet is true?

A. It differs from a cutout box in that the cover is removable.
B. It is intended to be used with switchboards.
C. It cannot be installed in a dwelling unit.
D. It cannot be installed in wet locations.

ANSWER ______________

PROCEDURE TO ESTABLISH ANSWER

The question is about a cabinet.

1. In Index, find "Cabinets, cutout boxes, and meter socket enclosures," under which, find "Definitions, Art. 100."

2. Review the definitions of *cabinet* and of *cutout box*. The definitions indicate that the cover of a cutout box is not removable, whereas the cover of a cabinet is removable. This establishes that Answer A is true.

3. Review the definition of *switchboard* and find that switchboards are not intended to be installed in cabinets. Answer B is not true.

4. Section 312.2 provides that enclosures covered in the article are permitted to be in damp or wet locations under the prescribed conditions. Answer D is not true.

5. Scanning Article 312, you find there is no prohibition for installing cabinets in dwelling units. Answer C is not true.

6. The correct answer is A.

18

A cutout box may be installed:

A. Flush mounted if at least 1.8 m (6 ft) from the floor
B. With removable covers
C. Recessed if used for a panelboard
D. Surface-mounted only

ANSWER ______________

PROCEDURE TO ESTABLISH ANSWER

The question is about cutout boxes.

1. In Index, find "Cutout boxes," under which, find "Definition, Art. 100."

2. In Article 100, Definitions, find *cutout box*.

3. The definition indicates that a cutout box is designed for surface mounting only.

4. The correct answer is D.

19

Which one of the following statements about the arrangement of the handles of circuit breaker type overcurrent protective devices in a switchboard is true?

A. Where the handles are operated horizontally, the "right" position must be the "on" position.
B. Where the handles are operated rotationally, the "down" position must be the "on" position.
C. Where the handles are operated vertically, the "up" position must be the "on" position.
D. The "up" position may be the "off" position if the switchboard is tested and identified with this arrangement.

ANSWER ______________

PROCEDURE TO ESTABLISH ANSWER

The question is about overcurrent protective devices, specifically, the position of circuit breaker handles.

1. In Index, find "Overcurrent protection, Art. 240," under which find "Circuit breakers, Part 240– VII."

2. Scan Article 240, Part VII, and find "240.81, Indicating," that requires that the "up" position of a circuit breaker be the "on" position where the handles are operated vertically. There is no provision for permitting the "up" position to be the "off" position. Also, there is no requirement relative to "on" position where the handle is operated horizontally or rotationally.

3. The correct answer is C.

20

Metal cabinets and boxes protected from corrosion solely by enamel are permitted to be used in which one of the following locations?

A. In outdoor locations
B. In car washes
C. In dairies
D. In enclosed parking garages

ANSWER ______________

PROCEDURE TO ESTABLISH ANSWER

The question is about protection against corrosion of cabinets.

1. In Index, find "Cabinets, cutout boxes, and meter socket enclosures," under which none of the subtitles appear applicable.

2. In Index, find "Protection," under which, find "Corrosion," under which find "Metal equipment, 300.6(A) and (B), 312.10(A)."

3. Section 300.6(A)(1) provides that, where protected against corrosion solely by enamel, cabinets are not to be used out-of-doors or in wet locations. The locations listed in answers A, B and C would be considered to be wet locations. The enclosed parking garage is assumed to be a damp or dry location. (See the definitions of "Locations wet" and Locations damp" in Article 100")

4. The correct answer is D.

21

Which one of the following statements about a cabinet mounted in a damp location is **NOT** true?

A. A metal cabinet must be mounted so there is at least 6 mm (¼ in.) of airspace between the cabinet and a concrete wall.
B. A nonmetallic cabinet must be mounted so there is at least 6 mm (¼ in.) of airspace between the cabinet and a concrete wall.
C. A metal cabinet must be mounted so there is at least 6 mm (¼ in.) of airspace between the cabinet and a wood wall.
D. A nonmetallic cabinet must be mounted so there is at least 6 mm (¼ in.) of airspace between the cabinet and a wood wall.

ANSWER ______________

PROCEDURE TO ESTABLISH ANSWER

The question is about cabinets and cutout boxes located in damp locations.

1. The question is about cabinets and cutout boxes located in damp locations.

2. Section 312.2 requires 6 mm (¼ in.) of airspace between a metal cabinet and walls regardless of the wall construction. Nonmetallic cabinets also require the 6 mm (¼ in.) airspace except where mounted on concrete, masonry, tile, or similar walls where no airspace is required.

3. The correct answer is B.

22

Which one of the following statements regarding cabinets recessed into walls is true?

A. The front edge of a cabinet recessed into a concrete wall may be set back of the finished surface up to 13 mm (½ in.).
B. The front edge of a cabinet recessed into a wall constructed of wood may be set back of the finished surface up to 6 mm (¼ in.).
C. The front edge of a cabinet recessed into a wall constructed of wood must be flush with the finished surface or project out from the surface.
D. The front edge of a cabinet must be flush with or project out from the finished surface of walls regardless of the wall construction.

ANSWER ______________

PROCEDURE TO ESTABLISH ANSWER

The question is about the position of cabinets recessed into walls.

1. In Index, find "Cabinets, cutout boxes, and meter socket enclosures," under which, find "Position in walls, 312.3."

2. Section 312.3 requires that a cabinet mounted in a finished wall of a concrete or noncombustible material shall be set back no more than 6 mm (¼ in.). In walls constructed of wood or other combustible material, the cabinet must be flush with the finished surface or project therefrom.

3. The correct answer is C.

23

A panelboard has four single-pole 20-ampere circuit breakers with neutral connections used for continuous lighting; four two-pole breakers with no neutral connections for heating load; and ten three-pole breakers for motor loads. There is a main breaker installed. Which one of the following statements is true?

A. This panelboard is defined as a lighting and appliance branch-circuit panelboard.
B. The 20-ampere breakers can be continuously loaded only to 16 amperes each.
C. This panelboard cannot contain more than 42 overcurrent devices.
D. This panelboard is considered to contain 22 overcurrent devices.

ANSWER ______________

PROCEDURE TO ESTABLISH ANSWER

The question is definition and number of circuit breakers in panelboards.

1. In Index, find "Panelboards," under which, you will not find "Lighting and appliance branch-circuit panelboards," as the term was deleted from the *NEC* in the 2008 edition. Answer A is not true.

   Note: There is a definition in Article 100-Part I, of a *panelboard* only.

2. Twenty ampere circuit breakers may carry continuous loads up to 16 amperes, maximum, per 210.19(A)(1).

   16 amperes x 125% = 20 amperes
   20 amperes x .80 = 16 amperes.

   This coincides with the UL listing for standard circuit breakers, which limits continuous loads on standard circuit breakers to 80% of their rating. Only circuit breakers listed for continuous use at 100% of the rating may be loaded to 100% when used in equipment that is designed for 100% continuous loading. Answer B is true as there is no mention of the circuit breaker or equipment being rated for 100% continuous loading.

3. Section 408.54 states that a panelboard must contain physical means to prevent the installation of more overcurrent devices than the panelboard was designed for. The same section also states that a 2-pole breaker is considered two overcurrent devices and a 3-pole breaker is three overcurrent devices. Considering this information, and the number of overcurrent devices given in the question, there would be 42 circuits in this panelboard. The total is four 1-pole @ 1 each; plus four 2-pole @ 2 each; plus ten 3-pole @ 3 each, or 4 poles + 8 poles + 30 poles for a total of 42 overcurrent devices. We find that answer D is not true.

4. Where a main breaker is installed within or in the supply circuit located ahead of the panelboard, Section 408.36 does not limit the number of these devices to 42. We find that answer D is not true.

5. The correct answer is B.

   Note: Section 210.20(A) limits the total load on an overcurrent device to not less than the noncontinuous load plus 125% of the continuous load. In this case, the load is continuous, and 16 amperes x 125% = 20 amperes. This is the equivalent of a maximum load on the breaker of not more than 80% of its rating.

24 The minimum wire bending space required for deflection of one 750-kcmil cable from the end of the lugs to the wall of the cabinet where the conductors enter or leave the cabinet opposite the terminal is:

A. 203 mm (8 in.)
B. 305 mm (12 in.)
C. 356 mm (14 in.)
D. 432 mm (17 in.)

ANSWER ____________

PROCEDURE TO ESTABLISH ANSWER

The question is about wire-bending space in a cabinet.

1. In Index, find "Cabinets, cutout boxes, and meter socket enclosures," under which, find "Wire bending space at terminals, 312.6(B)."

2. Section 312.6 has two tables, Tables 312.6(A) and 312.6(B). To determine where each table applies, refer to Sections 312.6(B)(1) and (2). Also note the exceptions.

3. Since the question applies to where the conductor enters or leaves the cabinet opposite the lug, Section 312.6(B)(2) requires the use of Table 312.6(B).

4. Refer to Table 312.6(B) and find that under wire size 750 kcmil, one conductor per terminal requires 432 mm (17 in.). The question does not mention a lay-in terminal; therefore, Note 2 under the table and the superscript letter "e" do not apply.

5. The correct answer is D.

25

Which one of the following statements about the mounting of a knife switch is true?

A. Where the throw of a double-throw switch is vertical, it must be able to be locked to hold the blades in the open position when so set.
B. Double-throw switches must be mounted so the throw is always in the vertical position.
C. Where the throw is horizontal, the switch must be able to be locked in either position when so set.
D. Single-throw switches must always be placed so that the switch contacts are energized when the switch is in the open position.

ANSWER ______________

PROCEDURE TO ESTABLISH ANSWER

The question is about the position of knife switches.

1. In Index, find "Knife switches," under which, find "Position, enclosed and open types, 404.6."

2. Section 404.6(B), Double-Throw Knife Switches, states double-throw knife switches may be mounted either vertically or horizontally. Locking is not required when the throw is horizontal. Answer B is not true.

3. Section 404.6(B), Double-Throw Knife Switches, also states that where the throw is vertical, a locking device shall be provided to hold the blades in the open position when so set. Double-throw knife switches may be mounted either vertically or horizontally. Locking is not required when the throw is horizontal. Answer A is true.

4. Section 404.6(A) permits single-throw knife switches to be mounted in the inverted position if provided with a locking device to ensure the blades remain in the open position when so set. Answer C is not true.

   Section 404.6(C) provides that single-throw knife switches be connected so the blades are de-energized when the switch is in the open position. Answer D could be an electrical hazard, if contacted by personnel. Answer D is not true.

5. The correct answer is A.

26

A main circuit breaker installed inside a cabinet must have its center of the grip of the operating handle where in the highest position not more than how far above a floor?

A. 1.8 m (6 ft)
B. 1.98 m (6 ft 6 in.)
C. 2.0 m (6 ft 7 in.)
D. 2.5 m (8 ft)

ANSWER ____________

PROCEDURE TO ESTABLISH ANSWER

The question is about the height of circuit breaker handles.

1. In Index, find "Circuit breakers," under which, find "Accessibility and grouping, 404.8(A)."

2. Section 404.8(A), Location, requires the circuit breaker to be so installed that the center of the grip of the operating handle when in its highest position will not be more than 2.0 m (6 ft 7 in.) above the floor or working platform.

3. The correct answer is C.

27

Which of the following statements about non-metallic sheathed cables or conductors entering a panelboard cabinet are true?

A. Conductors entering the panelboard must be protected from abrasion.
B. Several cables may be fed through a metric designator 53 (2 in.) raceway into the top of the cabinet if secured within 12 inches of where they leave the raceway.
C. Openings through which conductors enter shall be adequately closed.
D. All of the above.

ANSWER ____________

PROCEDURE TO ESTABLISH ANSWER

The question is about conductors and cables entering a panelboard cabinet.

1. In Index, find "Conductors," under which, find "Cabinets and cutout boxes, 312.5 through 312.7."

2. Section 312.5 requires conductors to be protected against abrasion, and subsection (A) requires openings through which conductors enter to be adequately closed. Answers A and C are true.

3. Unless the method described in the exception is used, 312.5(C) requires each cable to be secured to the cabinet.

4. Section 312.5(C), Exception, permits cables which are nonmetallic-sheathed to enter the top of a surface-mounted enclosure through one or more nonflexible raceways not less than 450 mm (18 in.) or more than 3.0 m (10 ft) in length with specific conditions in (1) through (7) to be followed.

5. Items in Answers A, B and C are all mentioned in the exception.

6. The correct answer is D.

28 Which one of the following statements regarding conductors that pass through a cabinet containing overcurrent devices is true?

A. They may fill up to 40 percent of the cross-sectional area of the wiring space.
B. They may fill up to 75 percent of the cross-sectional area of the gutter space.
C. There shall be no splices or taps in the conductors.
D. They are not permitted to feed through one cabinet to supply another cabinet.

ANSWER ______________

PROCEDURE TO ESTABLISH ANSWER

The question is about wiring, splices or taps in cabinets containing overcurrent devices.

1. In Index, find "Cabinets, cutout boxes, and meter socket enclosures," under which, find "Switch enclosures, splices and taps, 312.8."

2. Section 312.8(A) provides that enclosures for overcurrent devices can be used to splice or as a gutter or raceway for conductors feeding through or tapping off provided there is adequate space. Answers C and D are not true.

3. Conductors may fill the wiring space up to 40 percent of the cross-sectional area of the space, and conductors, splices and taps may fill the space up to 75 percent. Conductors alone may not fill the wiring space — which is equivalent to gutter space — more than 40%.

4. The correct answer is A.

29 Two 500-kcmil conductors per terminal enter or leave the top of a 400-ampere rated panelboard opposite their terminals. No conductors terminate in the bottom of the enclosure. The panelboard shall have a minimum top and bottom wire bending space of:

A. 356 mm (14 in.) of wire bending space at the top and bottom of the panelboard
B. 203 mm (8 in.) of wire bending space at the top and bottom of the panelboard
C. 203 mm (8 in.) of wire bending space at the top and 356 mm (14 in.) of wire bending space at the bottom of the panelboard
D. 356 mm (14 in.) of wire bending space at the top and 203 mm (8 in.) of wire bending space at the bottom of the panelboard

ANSWER ______________

PROCEDURE TO ESTABLISH ANSWER

The question is about wire bending space in panelboards.

1. In Index, find "Panelboards," under which, find "Wire bending space, 408.55."

2. Section 408.55 states the enclosure for a panelboard shall have the top and bottom bending space size in accordance with Table 312.6(B), etc.

3. Section 408.55, Exception No. 4, however, states either the top or bottom bending space, but not both, is permitted to be sized according to Table 312.6(A) where there are no conductors terminated in that space.

4. For two 500-kcmil conductors per terminal, Table 312.6(B) requires 356 mm (14 in.) bending space for top of the panelboard, while Table 312.6(A) requires 203 mm (8 in.) for the bottom since no conductors terminate in the bottom of the enclosure.

5. The correct answer is D.

30

Two 750-kcmil aluminum conductors per terminal (removable or "lay-in" type) enter an 800-ampere panelboard through the wall opposite its terminal. What is the minimum wire bending space permitted from the terminal to the opposite wall?

A. 356 mm (14 in.)
B. 406 mm (16 in.)
C. 432 mm (17 in.)
D. 483 mm (19 in.)

ANSWER _______________

PROCEDURE TO ESTABLISH ANSWER

The question is about wire bending space in panelboards.

1. In Index, find "Panelboards," under which, find "Wire bending space, 408.55."

2. Section 408.55 requires the bending space to comply with Table 312.6(B) with several exceptions. Since none of the exceptions apply, refer to Table 312.6(B).

3. Table 312.6(B) requires a 483 mm (19 in.) bending space for two 750-kcmil conductors.

4. Note the superscript "e" beside the dimensions, which refers to the notes at the bottom of the table.

5. Since the terminals are the lay-in type, Note 2 applies. Referring to the note below Table 312.6(B), find that the bending space may be reduced by 76.2 mm (3 in.) for those sizes indicated by the superscript "e."
   483 mm (19 in.) – 76.2 mm (3 in.) = 406 mm (16 in.)

6. The correct answer is B.

31

In an existing factory, a 277/480-volt switchboard faces a 120/240-volt panelboard on an opposite wall. If the switchboard is being replaced, and there are qualified persons servicing the installation, and written procedures prevent opening of both enclosures at the same time, what is the minimum working space between the new switchboard and the panelboard?

A. 762 mm (30 in.)
B. 914 mm (36 in.)
C. 1.07 m (3½ ft)
D. 1.2 m (4 ft)

ANSWER _______________

PROCEDURE TO ESTABLISH ANSWER

The question is about working space in front of a switchboard.

1. In Index, find "Working Space," under which, find "About Electrical Equipment, 110.26, 110.32 through 110.34, etc."

2. Section 110.26(A)(1) requires the working space to be not less than indicated in Table 110.26(A)(1).

   The table clearances are based on voltage and three conditions. Where electrical equipment is mounted opposite each other, it is assumed that there may be exposed live parts on both sides of the workspace and condition 3 applies.

3. Since the switchboard operates at 277 volts-to-ground, the 151– 600-volt clearance applies and condition 3 would require a 1.2 m (4 ft) clearance.

4. However, paragraph (c) under Table 110.26(A)(1), allows for replacements in existing buildings with qualified persons and written safety procedures that prevent opening of both enclosures at both times to comply with condition 2, which is 1.07 m (3½ ft).

5. The correct answer is C.

32

What is the minimum working space permitted in front of a panelboard where the voltage to ground is 120 volts?

A. 914 mm (36 in.) wide and 914 mm (36 in.) directly in front
B. 762 mm (30 in.) wide and 914 mm (36 in.) directly in front
C. 762 mm (30 in.) wide and 825 mm (33 in.) directly in front
D. 610 mm (24 in.) wide and 610 mm (24 in.) directly in front

ANSWER ______________

PROCEDURE TO ESTABLISH ANSWER

The question is about working space in front of panelboards.

1. In Index, find "Working Space," under which, find "About Electrical Equipment, 110.26, 110.32 through 110.34, etc."

2. Section 110.26(A)(2) requires the workspace to be not less than 762 mm (30 in.) wide.

3. Table 110.26(A)(1) requires 914 mm (36 in.) directly in front of the panelboard.

4. The correct answer is B.

33

What is the minimum depth of an outlet box without enclosed devices or utilization equipment?

A. 12.7 mm (½ in.)
B. 23.8 mm (15/16 in.)
C. 31.8 mm (1¼ in.)
D. 38.0 mm (1½ in.)

ANSWER ______________

PROCEDURE TO ESTABLISH ANSWER

The question is about the depth of outlet boxes.

1. In Index, find "Boxes (outlet, device, pull, and junction)" under which, find "Depth, minimum 314.24."
2. Section 314.24(A) requires a box that does not enclose devices or utilization equipment to have an internal depth of not less than 12.7 mm (½ in.).
3. The correct answer is A.

34

A metal wireway is installed in a factory between a panelboard and a machine. Which one of the following statements about the number of conductors in the wireway is true?

A. The wireway is not permitted to carry more than 30 conductors under any condition.
B. The wireway may contain more than 30 current-carrying conductors if adjusted (derated) for the number of current-carrying conductors and limited to not more than 30 percent fill.
C. Conductors shall have the ampacity adjusted where installed in a wireway and there are more than 30 current-carrying conductors at any cross-section.
D. The wireway may be filled to 40 percent of its area if the conductors are derated for conductor fill.

ANSWER _______________

PROCEDURE TO ESTABLISH ANSWER

The question is about conductor fill for wireways.

1. In Index, find "Wireways, metal" under which find, "Number of conductors, 376.22."
2. Section 376.22(B) permits a wireway to contain more than 30 current-carrying conductors, if adjusted (derated) for the number of current-carrying conductors according to Section 310.15(C)(1) and Table 310.15(C)(1). Answer A is not true. Answer C is true. In accordance with 376.22(A), the cross-sectional area of the conductors cannot exceed 20 percent of the area of the wireway at any cross-section. Answers B and D are not true.
3. The correct answer is C.

35

Where installing 3/0 AWG conductors in a metric designator 53 (Size 2) electrical metallic tubing that enter the back of a junction box opposite a removable cover and then are bent down to leave the bottom of the box. What is the minimum permitted depth of this junction box?

A. 102 mm (4 in.)
B. 165 mm (6½ in.)
C. 305 mm (12 in.)
D. 406 mm (16 in.)

ANSWER _______________

PROCEDURE TO ESTABLISH ANSWER

The question is about the size of a junction box.

1. In Index, find "Junction boxes," under which, find "Size," under which, find "Conductors 4 AWG and larger, 314.16, 314.28(A)."

2. Scan 314.28 and find (A)(2), "Angle or U Pulls or Splices," that requires angle pulls with the distance between raceway entry and the opposite wall of the box to be not less than six times the metric designator (trade size) of the largest raceway entry.

3. For a metric designator 53 (2 in.) tubing, this would require a 305 mm (12 in.) depth.

4. Note, however, that the exception to 314.28(A)(2) provides that where the raceway entry is in a wall opposite to a removable cover, the distance from that wall to the cover may be in accordance with the column for one wire per terminal in Table 312.6(A).

5. Reviewing Table 312.6(A), you will find that for one 3/0 AWG conductor per terminal, a 102 mm (4 in.) depth would be permitted.

6. The correct answer is A.

36

When determining the number of conductors in an outlet box, which one of the following statements is true?

A. Each equipment grounding conductor entering the box counts as one conductor volume.
B. Each internal cable clamp in the box counts as one conductor volume.
C. A cable connector clamping on the outside of the box counts as one conductor volume.
D. A receptacle in the box counts as two conductor volumes.

ANSWER ______________

PROCEDURE TO ESTABLISH ANSWER

The question is about box fill in outlet boxes.

1. In Index, find "Boxes (outlet, device, pull, and junction)" under which, find "Conductors, number in box, 314.16."

2. Section 314.16 provides information for box fill of conductors.

3. Section 314.16(B)(2) states that only one conductor is counted for all internal cable clamps, while no allowance is required for cable connectors clamping on the outside of the box. Answers B and C are not true.

4. Section 314.16(B)(5) requires counting only one conductor for all equipment grounding conductors entering the box. Answer A is not true.

5. Section 314.16(B)(4) requires counting two conductors for each device, such as a receptacle. Answer D is true.

6. The correct answer is D.

37 In a new installation, which one of the following statements about grounding in a branch-circuit panelboard remotely located from the service disconnecting means enclosure is true?

A. The panelboard cabinet may be grounded by a bonding screw or strap through the neutral bus to the enclosure.
B. The branch-circuit equipment grounding conductors may be connected to the neutral bus.
C. The neutral bus shall be connected to the cold-water pipe.
D. The neutral bus shall be isolated from the cabinet enclosure.

ANSWER ______________

PROCEDURE TO ESTABLISH ANSWER

The question is about methods of grounding in a branch-circuit panelboard, which is on the load side and remote from the service disconnecting means.

1. In Index, find "Grounding," under which, find "Panelboards, 408.40, 517.19(E)."

2. Section 408.40 requires panelboard cabinets to be grounded in accordance with Article 250.

3. Section 250.1 provides information on what is covered in each part of Article 250. Article 250, Part VII covers equipment grounding. Since most of the question involves grounding to the neutral, scan Article 250 Part VII and find "250.142, Use of Grounded Circuit Conductor for Grounding Equipment" and "(B), Load-Side Equipment," that provides generally that a grounded circuit conductor (neutral) is not to be used for grounding equipment on the load side of the service disconnecting means. Answers A and B are not true and statement D is true.

4. Knowledge of Article 250 leads us to 250.24(A), which provides that the grounding electrode conductor from a grounding electrode is connected to the grounded circuit conductor (neutral) and, in this case, between the load end of the service drop or lateral and the service disconnecting means as described in 250.24(A)(1).

5. There is no permission to ground the neutral in the branch-circuit panelboard to a grounding electrode (as in a cold-water pipe in contact with the earth for 3 m (10 ft.) or more) under the conditions stated in the question. Answer C is not true.

6. The correct answer is D.

38 Which one of the following statements regarding conductors that pass through a cabinet that contains overcurrent devices is NOT true?

A. The panelboard may be used as a wireway or junction box where adequate space is provided.
B. The panelboard must have sufficient space to accommodate all conductors installed in them without overcrowding.
C. Taps may fill the panelboard gutter to a maximum of 65 percent of the gutter area.
D. Conductors may fill the panelboard gutter up to 40 percent of the cross-sectional area of the gutter.

ANSWER ______________

PROCEDURE TO ESTABLISH ANSWER

The question is about conductors passing through cabinets.

1. In Index, find "Cabinets, cutout boxes, and meter socket enclosures," under which, find "Switch enclosures, splices, taps, 312.8."

2. Section 312.8 permits a cabinet containing overcurrent devices to be used as junction boxes, auxiliary gutters, or wireways, where adequate space is provided. Answers A and B are true.

3. Section 312.8 also permits conductor fill to 40 percent and 75 percent for conductors, splices and taps. Answer D is true; Answer C is not true.

4. The correct answer is C.

39

What is the minimum thickness permitted for a sheet metal steel of an uncoated electrical panelboard cabinet?

A. 1.35 mm (0.053 in.)
B. 1.52 mm (0.060 in.)
C. 1.59 mm (0.063 in.)
D. 2.29 mm (0.090 in.)

ANSWER _______________

PROCEDURE TO ESTABLISH ANSWER

The question is about the material from which cabinets and cutout boxes are constructed.

1. In Index, find "Cabinets, cutout boxes, and meter socket enclosures," under which, find "Construction specifications, 312– II."

2. Scan Part II of Article 312 and find "312.10(B)," that limits sheet steel cabinet construction to a minimum of 1.35 mm (0.053 in.) uncoated.

3. The correct answer is A.

40

What is the minimum length of a junction box in a straight pull of a metric designator 41 (Trade Size 1½) rigid metal conduit containing three 1/0 AWG Type THW conductors without splices?

A. 150 mm (6 in.)
B. 200 mm (8 in.)
C. 250 mm (10 in.)
D. 300 mm (12 in.)

ANSWER _______________

PROCEDURE TO ESTABLISH ANSWER

The question is about the size of a junction box.

1. In Index, find "Junction boxes," under which, find "Size," under which, find "Conductors 4 AWG and

larger, 314.16, 314.28(A)."

2. Section 314.28(A)(1) provides that in straight pulls, the length of the box must not be less than eight times the metric designator (trade size) of the largest raceway.

3. Eight times metric designator 41 (Trade Size 1 ½) requires a minimum length of 305 mm (12 in.).

4. The correct answer is D.

Note: The note to Table 300.1(C) states that the metric designators and trade sizes are for identification purposes only and are not actual dimensions.

41

Which of the following statements about the use of a raceway through which cables enter the top of a a surface mounted panelboard cabinet without being secured to the cabinet is NOT true?

A. The raceway is required to be of the nonflexible type.
B. The raceway is permitted to penetrate a structural ceiling.
C. The raceway must be sealed or plugged at the outer end in an approved manner.
D. The raceway must be secured at the outer end and at other points in accordance with the applicable article.

ANSWER ______________

PROCEDURE TO ESTABLISH ANSWER

The question is about rules under which it is not required to secure cables to panelboard cabinets.

1. In Index, find "Cabinets, cutout boxes, and meter socket enclosures, Art. 312."

2. Since there is no direct entry in the index, turn to Article 312.

3. Scan Article 312 and find "312.5(C), Cables."

4. The opening paragraph of 312.5(C), Exception, requires the use of nonflexible raceways. Answer A is true.

5. (2) to the exception does not permit the raceway to penetrate a structural ceiling. Answer B is not true.

6. (4) to the exception requires the outer end to be sealed or plugged in an approved manner. Answer C is true.

7. (6) to the exception requires the raceway to be fastened. Answer D is true.

8. The correct answer is B.

42 All statements about circuit identification in panelboard directories are true EXCEPT:

A. Circuits must be legibly identified as to the clear purpose or use with sufficient detail to distinguish each circuit from all other circuits.
B. Every circuit and circuit modification must be legibly identified.
C. Square meters (or sq. ft) for the area served must be listed.
D. Circuit identification is required to be located on the face or inside door of the panelboard.

ANSWER ______________

PROCEDURE TO ESTABLISH ANSWER

This question is about identification of circuits in panelboards.

1. In Index, find "Panelboards, Art. 408," under which find "Circuit directory, 408.4."

2. In 408.4(A), we find answers A, B, and D to be true.

3. The square meters (or sq. ft) area is not required to be listed in the directory. Answer C is not true.

4. The correct answer is C.

43 All are true about cabinet installations EXCEPT:

A. Metal cabinets installed in damp locations shall be mounted so that there is at least 6-mm (¼-in.) of air space between the enclosure and the wall or supporting surface.
B. Cabinets installed in damp locations shall be mounted so that there is at least 12-mm (½-in.) of air space between the enclosure and the wall or supporting surface.
C. If walls are constructed of wood or other combustible material, cabinets shall be flush with the finished surface, or project out from it.
D. Noncombustible (such as plaster, drywall, plasterboard, etc.) surfaces that are broken or incomplete must be repaired so that there will be no gaps or open spaces more than 3 mm (1/8 in.) at the edge of a cabinet or cutout box employing a flush-type cover.

ANSWER ______________

PROCEDURE TO ESTABLISH ANSWER

The question is about the installation of cabinets.

1. In Index, find "Cabinets, cutout boxes and meter socket enclosures, Art. 312," under which, find "Damp and wet locations 312.2."

2. Section 312.2 Damp and Wet Locations requires metal cabinets to be mounted so that there is at least 6-mm (¼-in.) air space between the enclosure and the wall or supporting surface. Answer A is true. Answer B is not true.

3. Section 312.3, Position in Wall, states, "In walls constructed of wood or other combustible material, cabinets shall be flush with the finished surface or project therefrom." Answer C is true.

4. Section 312.4, Repairing Noncombustible Surfaces, states that "…surfaces that are broken or incomplete

shall be repaired so that there will be no gaps or open spaces greater than 3 mm (1/8 in.) at the edge of the cabinet or cutout box employing a flush-type cover." Answer D is true.

5. The correct answer is B.

44 Which of the following statements is NOT true regarding metal-enclosed power switchgear?

A. Interrupter switches with stored energy mechanisms shall have mechanical interlocks to prevent access to the switch compartment unless the stored energy mechanism is discharged.

B. Energized bare parts mounted on doors shall be guarded where the door must be opened for maintenance of equipment or removal of draw-out equipment.

C. Gas discharged during operation of interrupting devices shall be directed so as to not endanger personnel.

D. The grounding bus installed in high-voltage service equipment shall not be extended into the compartment or section where the service conductors terminate.

ANSWER _______________

PROCEDURE TO ESTABLISH ANSWER

The question is about metal-enclosed power switchgear.

1. In Index, find "Metal-enclosed switchgear," which reads "see Switchgear, metal-enclosed."

2. Again, in the Index, find "Switchgear, metal-enclosed, 110.33(A), 110.34(F), 230.211, 250.112(A), 490–III."

3. In Article 490, scan Part III, and find "490.33, Guarding of Energized Parts Operating at 1000 Volts, Nominal, or Less Within Compartments," where we find that Answer B is true.

4. Further scanning in 490, Part III, we find "490.39," which verifies that Answer C is true.

5. Scanning further in 490, Part III, we find "490.42," which verifies that Answer A is true.

6. Finally, as we scan one last time, we find in 490, Part III, "490. 47," that requires the ground bus installed in metal-clad and metal-enclosed switchgear to be extended into the compartment in the switchgear where the service conductors terminate. Answer D is not true.

7. Answer D is correct.

CHAPTER 6
CONDUCTORS

1 A metal wireway contains forty-six conductors, with no more than ten conductors at any cross-sectional area of the wireway, providing less than a 20 percent cross-sectional area fill. Which one of the following statements about this installation is true?

A. The number of conductors could be increased to thirty at any cross-sectional area providing the fill did not exceed 20 percent.
B. The conductors are required to be adjusted (derated) factors based on Section 310.15(C)(1) .
C. The fill at any cross-sectional area may exceed 20 percent if the conductors are adjusted (derated) based on Section 310.15(C)(1) .
D. The number of conductors in the wireway is not permitted to exceed thirty.

ANSWER ______________

PROCEDURE TO ESTABLISH ANSWER

The question is about metal wireways.

1. In Index, find "Wireways, metal," under which, find "Number of conductors, 376.22."
2. Section 376.22(A) states the sum of the cross-sectional areas of all contained conductors at any cross section of a wireway shall not exceed 20 percent of the wireway. There is no limit to the number of conductors only to the percentage fill so Answer A is true, and Answers C and D are not true.
3. Section 376.22(B) further states that the adjustment (derating) factors in 310.15(C)(1) shall be applied only where the number of current-carrying conductors exceeds thirty. Answer B is not true as there were no more than 10 conductors in any cross-sectional area of the wireway.
4. The correct answer is A.

2 What is the maximum number of current-carrying conductors permitted within a nonmetallic wireway at any cross section?

A. 20
B. 30
C. 75
D. No limit

ANSWER ______________

PROCEDURE TO ESTABLISH ANSWER

The question is about number of conductors in a nonmetallic wireway.

1. In Index find "Wireways, nonmetallic," under which, find "Number of conductors, 378.22."

2. Section 378.22 does not specify any limit on the number of current-carrying conductors in a nonmetallic wireway but does limit the fill to not more than 20 percent of the cross-sectional area of the nonmetallic wireway. The plastic of the nonmetallic wireway will not absorb and act as a heat sink like the metallic wireway.
3. The correct answer is D.

   Note: The adjustment (derating) factors specified in Section 310.15(C)(1) apply where more than three current-carrying conductors are contained at any cross-sectional area of the nonmetallic wireway.

3 Which one of the following types of conductors is NOT acceptable for wiring in a fluorescent luminaire where the branch-circuit conductor is located within 75 mm (3 in.) of the ballast?

A. RHW
B. THWN-2
C. THHN
D. XHHW

ANSWER ______________

PROCEDURE TO ESTABLISH ANSWER

The question is about luminaires used as a raceway.

1. In Index, find "Luminaires," under which, find "Raceways, 410.30(B), 410.36(E), 410.64." The concern about luminaires, used as a raceway, is with the temperature of a conductor located near the ballast.
2. Section 410.64(A) specifies that luminaires shall not be used as a raceway for circuit conductors, unless listed and marked for use as a raceway.
3. Further scanning Article 410, you will find Section 410.68, which states that branch-circuit conductors within 75 mm (3 in.) of a ballast shall have an insulation temperature rating not lower than 90°C (194°F), unless supplying a luminaire marked as suitable for a different insulation temperature.
4. In Index, find "Conductors," under which, find "Insulated, Article 310, then go to Article 310" and locate Table 310.4(A). Insulated conductors are required to comply with Tables 310.4(A) and 310.4(B)."
5. Table 310.16 can also be used to identify insulation temperature ratings since conductor insulation temperature ratings are necessary for determining conductor ampacities.
6. In Table 310.4(A), note that conductors with types THWN-2, THHW, and XHHW insulation are 90°C rated. Answers B, C, and D are true. Type RHW conductors are 75°C rated. Answer A is not true.
7. The correct answer is A.

4 A 480-volt, 3-phase feeder of Type THW copper conductors are installed to serve the following five motors: two 25-horsepower, one 20-horsepower, one 7.5-horsepower and one 5-horsepower. The minimum required size of this feeder conductor is:

A. 2 AWG
B. 1 AWG
C. 1/0 AWG
D. 2/0 AWG

ANSWER _______________

PROCEDURE TO ESTABLISH ANSWER

The question is about the ampacity of conductors serving five motors.

1. In Index, find "Motors" under which find "General, 430–I." You can also find "Ampacities" in the Index under which find "Conductors" under which find "Motors, motor circuits and controllers, 430.6."

2. Scan Article 430, Part I, for applicable titles or look under 430.6 and find 430.6(A).

3. Section 430.6(A) generally requires the "full load current (FLC) rating" of a motor is to be used to determine the ampacity of conductors, etc. It specifies the values in Tables 430.247 through 430.250, including the notes, shall be used instead of the actual ampere (FLA) rating marked on the nameplate.

4. Table 430.250 is applicable; read down the horsepower column to the desired horsepower and across to the 460 volt column (used for the 440 V to 480 V range):

   Two 25-HP FLA of 34 A x 2 motors = 68.0 A
   One 20-HP FLA of 27 A x 1 motor = 27.0 A
   One 7.5-HP FLA of 11 A x 1 motor = 11.0 A
   One 5-HP FLA of 7.6 A x 1 motor = 7.6 A

   Total of all motors (68A + 27A +11.0A + 7.6A) = 113.6 amperes

5. In Index, find "Motors," under which, find "Conductors" under which find "Several motors, 430.24."

6. Section 430.24 requires conductors supplying several motors to have an ampacity equal to the sum of the full-load current ratings of all the motors (which is 113.6 amperes) plus 25 percent of the highest rated motor in the group:

   25% (largest motor) x 34 amperes = 8.5 amperes
   Total amperes of all motors plus 25% of full load amps of largest motor
   Total amperes =113.6 amperes + 8.5 amperes
   Total amperes =122.1 amperes

7. In Index, find "Conductors," under which, find "Ampacities, 310.14 through 310.21 , e."

8. Using Table 310.16 read down the Type THW column for copper to the next ampacity greater than 122.1 amperes and find that 130 amperes for a 1 AWG Type THW conductor required.

9. The correct answer is B.

5 Where determining the number of conductors to be installed in rigid metal conduit, a cable having an elliptical cross section shall be calculated using which part of the ellipse to determine cross-sectional area?

A. Minor diameter
B. Average diameter
C. Major diameter
D. Root Mean Square (R.M.S.)

ANSWER ______________

PROCEDURE TO ESTABLISH ANSWER

The question is about the number of conductors in rigid metal conduit.

1. In Index, find "Rigid metal conduit (Type RMC)," under which, find "Number of conductors, 344.22, Chap. 9, Table 1."

2. Section 344.22 refers to Chapter 9 Table 1.

3. Scan Table 1 and the Notes to Tables. Note 9 concerns cables and states that for cables which have an elliptical cross section, the cross-sectional area calculation shall be based on using the major diameter of the ellipse as a circle diameter for the formula for the area of a circle is $A = \pi r^2$ where "r" is the radius or $A = \pi (d)^2$ where "d" is the diameter

4. The correct answer is C.

6 Four 12 AWG Type THW and three 10 AWG Type THW conductors are being installed in a single rigid metal conduit. What is the minimum size rigid metal conduit required for this installation?

A. Metric designator 16 (½ in.)
B. Metric designator 21 (¾ in.)
C. Metric designator 27 (1 in.)
D. Metric designator 35 (1¼ in.)

ANSWER ______________

PROCEDURE TO ESTABLISH ANSWER

The question is about the minimum size of rigid metal conduit for specific conductors.

1. In Index, find "Conductor fill" under which find "Rigid metal conduit, 344.22."

2. Section 344.22 refers to Table 1, Chapter 9, which permits a 40 percent raceway fill for over two conductors in a raceway.

3. Note 6 to Table 1 states that, for combinations of conductors of different size, use the actual dimensions or Tables 5 and 5A in Chapter 9 for dimensions of conductors, and Table 4 for applicable conduit or tubing dimensions. Since no actual dimensions were provided in the question, we will use the tables.

4. Referring to Table 5 for conductor area,

12 AWG Type THW area of 0.0181 in^2 x (4) conductors	= 0.0724 in^2
10 AWG Type THW area of 0.0243 in^2 x (3) conductors	= 0.0729 in^2
Total area for seven conductors (0.0724 in^2 + 0.0729 in^2)	= 0.1453 sq. in.

5. Referring to Table 4 for "Rigid Metal Conduit" under the "Over 2 Wires 40%" column, a metric designator 16 (Trade Size ½) conduit permits a conductor fill of 0.125 sq. in., which is too small.

   A metric designator 21 (Trade Size ¾) conduit permits a conductor fill of 0.220 sq. in.

6. The correct answer is B.

7

In general, where the standard overcurrent device rating does not correspond with the ampacity rating of a conductor, the next higher overcurrent device rating may be used for which one of the following conditions?

A. The next higher standard rating exceeds 800 amperes.
B. The circuit conductors supply receptacles on a multioutlet circuit for cord-and-plug-connected loads.
C. The circuit only supplies 3-phase load.
D. The conductors are not part of a branch circuit supplying more than one receptacle cord-and-plug-connected portable loads and the next higher standard rating does not exceed 800 amperes.

ANSWER ______________

PROCEDURE TO ESTABLISH ANSWER

The question is about conductor overcurrent protection.

1. In Index, find "Conductors" under which find "Overcurrent protection, 240.4."

2. Scan Section 240.4(B), which confirms that Answer D is true. The code section states that the overcurrent device may not exceed 800 amps, nor supply receptacles on a multi-outlet branch circuit for cord-and-plug-connected portable loads but is silent regarding phase. A, B and C are not true.

   Also note some of the questions contain part of the requirement but do not satisfy all the conditions specified in the section, so special attention to the answers is necessary in questions like this.

3. The correct answer is D.

8

The minimum size main bonding jumper permitted for a 480/277-volt, 3-phase, 6000-ampere service installed in one switchboard enclosure, which is served by paralleling sixteen 500-kcmil copper conductors, is:

A. 1/0 AWG
B. 3/0 AWG
C. 500 kcmil
D. 1000 kcmil

ANSWER ______________

PROCEDURE TO ESTABLISH ANSWER

The question is about sizing of a main bonding jumper

1. In Index, find "Jumpers, bonding" under which find "Service equipment, 250.28, 250.92(B), 250.102(C), 250.186."

2. Section 250.28(D)(1) requires the main bonding jumper to be not smaller than shown in Table 250.102(C)(1) and Note 1 to the Table requires phase conductors larger than 1100 kcmil copper or 1750 kcmil aluminum to be sized not less than 12½ percent of the area of the larger phase conductor.

   In this case, we have parallel conductors with sixteen conductors per phase. As noted at the top of Table 250.102(C)(1), the size chosen for the bonding jumper is for a single conductor or the equivalent area for parallel conductors:

3. 16 conductors per phase x 500 kcmil = 8000 kcmil
   8000 kcmil x 12 ½% = 1000 kcmil

4. The correct answer is D.

9

What is the minimum size copper equipment grounding conductor permitted to be used for a circuit having 4/0 AWG phase conductors and protected with 250-ampere fuses?

A. 8 AWG
B. 6 AWG
C. 4 AWG
D. 3 AWG

ANSWER ______________

PROCEDURE TO ESTABLISH ANSWER

The question is about sizing of equipment grounding conductors.

1. In Index, find "Equipment grounding conductors," under which, find "Sizing, 250.122, 250.190(C)(3)."

2. Section 250.122 requires the equipment grounding conductor to be not less than shown in Table 250.122.

3. Since the 250-ampere fuse protection is between the 200- and 300-ampere rating it is necessary to base the

size of the equipment grounding conductor on the 300-ampere rating or a 4 AWG copper.

4. The correct answer is C.

10

The maximum distance permitted between supports for 750 kcmil aluminum conductors installed in a vertical raceway is:

A. 11 mm (35 ft)
B. 12 mm (40 ft)
C. 26 mm (85 ft)
D. 28 mm (95 ft)

ANSWER ______________

PROCEDURE TO ESTABLISH ANSWER

The question is about supporting requirements for conductors in a vertical raceway.

1. In Index, find "Conductors," under which, find "Support of, 110.12(C) , 300.19."

2. Section 300.19 refers to Table 300.19(A).

3. Table 300.19(A) shows 750 kcmil aluminum conductors requiring vertical support at 28 mm (95 ft) intervals.

4. The correct answer is D.

11

Unless specifically permitted or required to be otherwise, what is the largest size of solid conductor that can be pulled into a raceway where part of a multiwire branch-circuit?

A. 10 AWG
B. 8 AWG
C. 6 AWG
D. There are no restrictions.

ANSWER ______________

PROCEDURE TO ESTABLISH ANSWER

The question asks, *what is the largest size solid conductor that can be pulled into a raceway*? But it could also be looked at as *when are conductors required to be stranded where installed in a raceway ?* As we research the answer, we will find the *Code* looks at this from the "stranded" perspective.

1. In Index, find "Conductors," under which, find "Stranded, 311.12(C), Chapter 9, Table 10." Article 311 covers over 2000-volt conductors but Article 310 covers conductors that are 2000 volts and below so go to Article 310, Part I.

2. Section 310.3(C) states that conductors size 8 AWG and larger shall be stranded where installed in a raceway, unless permitted elsewhere in the *Code*. Therefore, the largest size solid permitted is a 10 AWG.
   Note: An example of where a solid conductor larger than 10 AWG is permitted in a raceway is 680.23(B)(2)(b) which permits an 8 AWG solid or stranded copper bonding jumper installed in a conduit to the forming shell of a wet-niche luminaire.

3. The correct answer is A.

12

Without considering any exceptions, which one of the following statements about a circuit consisting of both paralleled phase and neutral conductors is true?

A. Conductors larger than 1/0 AWG are not permitted to be connected in parallel.
B. Conductors in parallel are not subject to adjustment factors, as required in Section 310.15(C)(1) .
C. The phase conductors may be a different size than the neutral conductors.
D. Paralleled equipment grounding conductors run with this circuit are required to be a minimum size of 1/0 AWG.

ANSWER ______________

PROCEDURE TO ESTABLISH ANSWER

The question is about paralleled conductors.

1. In Index, find "Conductors" under which, find "Paralleled, 250.30(A)(3), 250.102(C), 250.122(F), 250.186(A)(2), 250.185(B)(2), 300.3(B)(1), 300.5(I) Ex. 1, 300.20, 310.10(G) , etc."

2. Section 310.10(G)(1) requires that phase and neutral conductors connected in parallel must be 1/0 AWG and larger. Answer A is not true.

3. Section 310.10(G)(4) states that conductors installed in parallel shall be subject to the ampacity reduction factors required by Section 310.15(C)(1) . Answer B is not true.

4. Section 310.10(G)(5) states equipment grounding conductors that are paralleled must comply with all of the requirements of Section 310.10(G)(1) through 310.10(G)(6) except that the equipment grounding conductors are to be sized according to Section 250.122 which would permit paralleled conductors smaller than 1/0 AWG. Answer D is not true. Equipment grounding conductors are sized based on the rating of the overcurrent device protecting the feeder or branch-circuit conductors in accordance with 250.122(F) and Table 250.122.

5. The correct answer is C.

   Note: The requirement for paralleled conductors to be the same size applies only to conductors of each phase, neutral circuit conductor, or equipment grounding conductors. The paralleled phase conductors may be a different size than the paralleled neutral conductors or equipment grounding conductors. Section 220.61, for example, permits a reduction in size of the neutral conductor, under certain conditions specified in that section. Therefore, Answer C is a true statement.

13

What is the smallest size service-entrance ungrounded conductor permitted to be paralleled?

A. 1 AWG
B. 1/0 AWG
C. 2/0 AWG
D. 3/0 AWG

ANSWER ______________

PROCEDURE TO ESTABLISH ANSWER

The question is about paralleled service-entrance conductors.

1. In Index, find "Conductors" under which, find "Paralleled, 250.30(A)(3), 250.102(C), 250.122(F), 250.186(A)(2), 250.185(B)(2), 300.3(B)(1), 300.5(I) Ex. 1, 300.20, 310.10(G) , etc."

2. Section 310.10(G)(1) covers all paralleled conductors 1/0 AWG and larger. There is no exception for service-entrance conductors.

3. The correct answer is B.

14

8 AWG Type THHN copper conductors are permitted to be installed in which one of the following locations?

A. Dry locations only
B. Dry and damp locations
C. Wet locations
D. Dry and wet locations

ANSWER ______________

PROCEDURE TO ESTABLISH ANSWER

The question is about the recognized use of insulation types.

1. In Index, find "Conductors" under which, find "Insulated, Article 310"."

2. Section 310.4 refers to Table 310.4(A).

3. Scan Table 310.4(A) and find Type THHN which is shown to be acceptable in dry and damp locations.

4. The correct answer is B.

15

A shopping center utilizes electric-discharge lighting supplied by 208Y/120-volt, 3-phase, 4-wire feeders consisting of 4/0 AWG Type THW copper conductors. What is the minimum size Type THW copper neutral conductor permitted for these feeders?

A. 1/0 AWG
B. 2/0 AWG
C. 3/0 AWG
D. 4/0 AWG

ANSWER ______________

PROCEDURE TO ESTABLISH ANSWER

The question is about sizing of neutral conductors for feeders supplying electric-discharge lighting.

1. In Index, find "Neutral" under which, find "Feeder and service load, 220.61, 220.82(A), 220.84(A), Annex D."

2. Section 220.61(A) states that the feeder neutral load is the maximum unbalance of the load, which is the

maximum load between the neutral and any one ungrounded conductor. Not knowing the specific load, it can be assumed that the 4/0 AWG conductors are necessary for the maximum load.

3. Section 220.61(C)(2) states that there shall be no reduction of the neutral capacity for that portion of the load that consists of nonlinear loads (such as fluorescent luminaires, electric-discharge lighting).

4. The correct answer is D.

16

Three Type THW copper conductors are feeding one bank of three 50 kVA, 480-volt capacitors connected to a 480-volt, 3-phase service. Which one of the following is the minimum size conductor permitted to supply this capacitor bank?

A. 1/0 AWG
B. 3/0 AWG
C. 250 kcmil
D. 350 kcmil

ANSWER ______________

PROCEDURE TO ESTABLISH ANSWER

The question is about the minimum size conductors for capacitor circuits.

1. In Index, find "Capacitors" under which, find "1000 volts and under, 460–I" under which, find "Conductors, 460.8."

2. Section 460.8(A) requires the capacitor circuit conductors to be not less than 135 percent of the rated current of the capacitor. (kVAr means thousands of volt-amperes reactive. We are talking capacitive reactance).

3. Determine the rated current of the capacitors as follows:

   50-kVAr x 3 = 150-kVAr
   150 kVAr x 1000 = 150,000-VAr

   I = VA/E x 1.73
   I = 150,000 VA/480 volts x 1.73
   I = 150,000 VA/830.4 V_{eff}
   I = 180.6 amperes

4. 181.6 A x 135% = 245.16 amperes.

5. In Index, find "Conductors" under which, find "Copper" under which, find "Ampacities, 310.14 through 310.21, etc."

6. Refer to Table 310.16 and read down the Type THW copper column to the first ampacity equal to or greater than 245 amperes.

7. Table 310.16 permits 255 amperes for a 250-kcmil conductor under the 75°C (THW) column. Size 4/0 is only rated for 230 amps.

8. The correct answer is C.

17 Three current-carrying Type THW feeder conductors in a raceway pass through the boiler room exposed on the ceiling where the temperature stays at 120°F. Which one of the following statements is true?

A. The conductors' ampacity must be reduced to 75 percent.
B. The conductors' ampacity must be reduced to 58 percent.
C. The conductors' ampacity must be reduced to 41 percent.
D. Derating of the conductors is not required.

ANSWER _______________

PROCEDURE TO ESTABLISH ANSWER

The question concerns the temperature limitations of conductors, therefore:

1. In Index, find "Conductors" under which, find "Temperature, limitations of, 310.14(A)(3) , 338.10(B)(3)."

2. Section 310.14(A)(3) states that conductors are not to be used in such a manner that the operating temperature of the conductor will exceed the rating of the insulation.

3. The Informational Note to refers to the ampacity tables and the correction factors in 310.15(B).

4. Referring to the correction factors in Table 310.15(B)(1), you will find Type THW (75°C) wire operating in an ambient of 120°F has an ambient temperature correction factor of .75.

5. The correct answer is A.

18 A conduit contains six Type THHN 8 AWG copper conductors. These branch circuits supply continuous loads in an industrial building. The conduit passes close to heat-producing equipment that subjects the conductors to an ambient temperature of 49°C. Which of the following is the maximum permitted ampere load on each conductor?

A. 32 amperes
B. 36 amperes
C. 40 amperes
D. 50 amperes

ANSWER _______________

PROCEDURE TO ESTABLISH ANSWER

The question is about the ampacity adjustment and correction (derating) of conductors and continuous loading of overcurrent devices. Although all three affect the ampacity of the circuit, the adjustment for the number of current-carrying conductors and the correction for the ambient temperature are separate considerations from the continuous loading on the overcurrent device. The adjustment and corrections are related to the use of the conductor, and continuous loading is related to the overcurrent device. We will establish a maximum permitted ampacity for each consideration and ultimately choose the lower value in this case.

1. In Index, find "Conductors" under which, find "Ampacities, 310.14 through 310.21."

2. Section 310.15 refers to Tables 310.16 through 310.21 and their accompanying notes.

3. Table 310.16 shows 8 AWG Type THHN at 55 amperes in a dry location in the 90°C column. However, because Table 310.16 is based on not more than three current-carrying conductors in a raceway and an ambient temperature of 30°C, adjustments for more than three current carrying conductor and ambient temperature corrections (derating) factors will be required to compensate for the additional current-carrying conductors and the higher ambient temperature.

4. Since the question did not provide any information on the temperature ratings of the terminals, 110.14(C)(1)(a) regarding temperature limitations will establish a 60°C temperature rating based on the fact that the stated conductor size, 8 AWG, is larger than 14 AWG and smaller than 1 AWG and the equipment is assumed be rated less than 100 amperes. However, the second sentence of 110.14(C)(1)(a)(3) that provides the higher temperature rating of the THHN insulation (90°C) can be used for ampacity adjustment or correction, so we can use 55 amperes for adjustment and correction.

5. In accordance with 310.15(B)(1) and based on the 49°C ambient and the 90°C temperature rating of the insulation, a correction factor of .82 is chosen from the 90°C column of Table 310.15(B)(1).

6. Based on the six current-carrying conductors in the conduit, 310.15(C)(1) and Table 310.15(C)(1) require an adjustment factor of 80 percent for 4 to 6 current-carrying conductors in a raceway.

7. The adjusted ampacity of the 8 AWG, THHN, conductor is:
   55 amperes x .82 multiplier = 45.1 amperes
   45.1 amperes x .80 = 36.08 amperes.

8. Section 210.20(A) states that the rating of the overcurrent device shall not be less than the noncontinuous load plus 125 percent of the continuous load or, in other words, limits the continuous load current to 80 percent of the overcurrent device rating.

9. As noted, 110.14(C)(1)(a) established a 60°C rating and there was no permission to use the higher temperature rating when considering continuous loading. Therefore, 40 amperes is chosen for this consideration.

   40 A x 80% amperes = 32 amperes which is the maximum permitted continuous load on a circuit breaker.

10. Because the value established for the continuous loading (32 amperes) is lower than the value established for the conditions of use of the conductors (36 amperes), the correct answer is A.

19 An outside 480Y/277-volt overhead feeder (open conductors) extends over a commercial area accessible to pedestrians but is not subject to truck traffic. What is the minimum clearance above grade?

A. 2.5 m (8 ft)
B. 3.0 m (10 ft)
C. 3.7 m (12 ft)
D. 4.5 m (15 ft)

ANSWER ______________

PROCEDURE TO ESTABLISH ANSWER

The question is about clearances of open feeder conductors above grade.

1. In Index, find "Clearances" under which, find "Conductors" under which, find "Open, outside branch circuits and feeders, 225.18, 225.19."

2. Section 225.18 "Clearances for Overhead Conductors and Cables," list item (2), requires the minimum clearance to be 3.7 m (12 ft) for conductors in commercial areas, not subject to truck traffic, where the supply voltage exceeds 150 volts to ground but does not exceed 300 volts to ground.

3. The correct answer is C.

20 Type TBS insulation is recognized for which one of the following applications?

A. Machine tool wiring
B. Appliance wiring
C. Fixture wiring
D. Switchboard wiring

ANSWER ______________

PROCEDURE TO ESTABLISH ANSWER

The question is about the recognized use of an insulation type.

1. In Index, find "Conductors, Article 310" under which find "Materials, 311.12(B) ." The TBS insulation is not used for more than 2000-Volt applications but go to Part I of Article 310.

2. Section 310.4 applies Table 310.4(A) to the types of insulation.

3. Scan Table 310.4(A) and find that Type TBS insulation is recognized for switchboard wiring only.

4. The correct answer is D.

21 What is the ampacity of a circuit consisting of three single direct-buried 4/0 AWG, 5000-volt copper conductors based on conductor temperature of 90°C?

A. 385 amperes
B. 405 amperes
C. 435 amperes
D. 470 amperes

ANSWER ______________

PROCEDURE TO ESTABLISH ANSWER

The question is about the ampacity of direct-buried 5000-volt copper conductors.

1. In Index, find "Conductors" under which, find "Ampacities, (Look in new Article 311), Annex B. Table B.2(1) through B.2(10) ."

2. Scan the ampacity tables in Article 311 for an ampacity table for single insulated conductors directly buried in earth that applies to the question above and find Table 311.60(C)(81) for the ampacities of 2001 through 35,000-volt conductors.

3. In Table 311.60(C)(81) , "Ampacities of Single Insulated Copper Conductors Directly Buried in Earth…," "2001–5000 Volts Ampacity, 90°C, Type MV-90, one circuit, three conductors," etc., and find that a 4/0 AWG copper conductor has an allowable ampacity of 435 amperes.

4. The correct answer is C.

22

The sizing of conductors supplying several motors is based on which one of the following?

A. Full-load current of all motors in the group
B. Full-load current of all motors plus 25 percent of highest rated motor in group
C. 125 percent of the full-load current of all motors in the group
D. 100 percent of the largest overcurrent device protecting the motors

ANSWER ______________

PROCEDURE TO ESTABLISH ANSWER

The question is about sizing motor circuit conductors.

1. In Index, find "Motors" under which, find "Conductors, 430–II" under which find "Several motors, 430.24."

2. Section 430.24 requires conductors supplying several motors to have an ampacity at least equal to the full-load current (FLC) rating of all the motors plus 25 percent of the highest rated motor in the group.

3. The correct answer is B.

23

Which one of the following cables is NOT permitted to be installed above the hazardous floor area in a major-repair type commercial garage that has ventilation?

A. Type UF
B. Type MC
C. Type AC
D. Type TC

ANSWER ______________

PROCEDURE TO ESTABLISH ANSWER

The question is about wiring methods in a commercial garage.

1. In Index, find "Commercial garages," which refers you to "See Garages, commercial" under which, find "Classification of locations, 511.3" and "Wiring, 511.4, 511.7."

2. Section 511.3(C) identifies that for areas where vehicles using Class 1 liquids or heavier-than-air fuels are repaired (both major and minor repairs), classification guidance for hazardous areas is found in Table 511.3(C). Table 511.3(C) classifies the area up to 450 mm (18 in.) above the floor as Class I, Division 2.

3. Section 511.7(A)(1) specifies the wiring methods permitted above the classified area. Types MC, AC, MI, and TC cables are permitted, while Type UF is not permitted.

4. The correct answer is A.

24 What adjustment factor (derating) is required to be applied to twenty current-carrying conductors installed in electrical nonmetallic tubing?

A. 35 percent
B. 40 percent
C. 50 percent
D. 70 percent

ANSWER ______________

PROCEDURE TO ESTABLISH ANSWER

The question is about ampacity reduction by adjustment for the number of conductors in electrical nonmetallic tubing where there will be more than three current-carrying conductors installed in the raceway.

1. In Index, find "Ampacities" under which, find "Conductors, 310.14 through 310.21, etc."

2. Section 310.15 refers to Tables 310.16 through 310.21.

3. Table 310.15(C)(1) provides adjustment factors to be applied when four or more current-carrying conductors are installed in a raceway.

4. Twenty conductors require an adjustment factor (multiplier) of 50 percent.

5. The correct answer is C.

25 A 3-wire circuit consisting of two-phase conductors and one neutral conductor is connected to a 208Y/120-volt, 4-wire system that supplies incandescent lighting. The load is equally divided with sixteen amperes on each of the two-phase conductors. What is the approximate current in the neutral conductor?

A. 0 amps
B. 8 amps
C. 16 amps
D. 32 amps

ANSWER ______________

PROCEDURE TO ESTABLISH ANSWER

The question is about current in the neutral conductor.

1. In Index, find "Neutral," under which, find "Conductor, 310.15(E), etc."

2. Section 310.15(E)(2) states that in a 3-wire circuit fed from a 4-wire, 3-phase wye-connected system, the common or neutral conductor carries approximately the same current as the line-to-neutral load currents of the other conductors.

3. The correct answer is C.

26

What is the minimum size rigid metal conduit required for two 2/0 AWG and two 1/0 AWG Type THW conductors?

A. Metric Designator 35 (1¼ in.)
B. Metric Designator 41 (1½ in.)
C. Metric Designator 53 (2 in.)
D. Metric Designator 63 (2½ in.)

ANSWER ______________

PROCEDURE TO ESTABLISH ANSWER

The question is about conductor fill in rigid metal conduit.

1. In Index, find "Conductor fill," under which, find "Rigid metal conduit, 344.22."

2. Section 344.22 refers to Table 1, Chapter 9, which permits a 40 percent raceway fill for over two conductors in a raceway.

3. Note 6 to Table 1 states that, for combinations of conductors of different sizes, use the actual size of the conductor or Tables 5 and 5A in Chapter 9 for dimensions of conductors and Table 4 for applicable conduit or tubing dimensions. Since no actual sizes were given in the question, we will use the Tables.

4. Referring to Table 5 for conductor area,
   2/0 AWG Type THW area of 0.2624 in^2 x 2 conductors = 0.5248 in^2
   1/0 AWG Type THW area of 0.2223 in^2 x 2 conductors = 0.4446 in^2
   Total area for 4 conductors (0.5248 in^2 + 0.4446 in^2) = 0.9694 sq. in.

5. Referring to Table 4 for "Rigid Metal Conduit," under the "Over 2 Wires 40%" column,

   A metric designator 41 (1½ in.) conduit permits a conductor fill of 0.829 sq. in.
   A metric designator 53 (2 in.) conduit permits a conductor fill of 1.363 sq. in.

6. The correct answer is C.

27

All of the following statements regarding requirements for conductors or cables exposed to direct sunlight are true EXCEPT:

A. Cables shall be listed or listed and marked "sunlight resistant."
B. Conductors shall be listed or listed and marked "sunlight resistant."
C. Conductors or cables shall be covered with insulating material, such as tape or sleeving that is listed, or listed and marked "sunlight resistant."
D. Cables or conductors listed for wet locations meet the requirements for sunlight resistance, where exposed to direct rays of the sun.

ANSWER ______________

PROCEDURE TO ESTABLISH ANSWER

The question is about sunlight resistance requirements for conductors exposed to direct sunlight.

1. Currently, there is no direct route to the answer in the Index. Attempts to find "Sunlight resistant" under "Sunlight resistant," "Conductors," "Cables," or "Protection" will not provide any direction. However, in the Index find "Conductors" under which find "Insulated, Article 310."

   You can also look in the Table of Contents, in the front of the book, and find Chapter 3, Wiring Methods. Since the conductors and cables being discussed are wiring methods, scan Chapter 3, Wiring Methods in the Table of Contents and find "Article 310, Conductors for General Wiring."

2. Scan Article 310, and find "310.10, Locations," under which, find "310.10(D), Locations Exposed to Direct Sunlight." Read 310.10(D)(1) and (2) and you'll find that answers A, B, and C are true. Answer D is not true.

3. The correct answer is D.

28

Where installed in raceways, conductors larger than what size are required to be stranded?

A. 10 AWG
B. 8 AWG
C. 6 AWG
D. 4 AWG

ANSWER ______________

PROCEDURE TO ESTABLISH ANSWER

The question is about installing conductors in raceways and is really asking "at what point is the conductor required to be stranded."

1. In Index, find "Conductors," under which, find "Stranded, 311.12(C) , Chapter 9, Table 10." Since 311.12(C) only applies to over 2000-volt applications, go to Part I of Article 310, scan the page, and find 310.3(C).

2. Section 310.3(C) requires 8 AWG and larger conductors, installed in raceways, to be stranded, unless specifically permitted or required elsewhere in this Code to be solid.

3. The correct answer is A.

29

Which one of the following conductors is NOT permitted to be installed in rigid polyvinyl chloride (PVC) conduit in a wet location of a car wash?

A. Type SA (silicone)
B. Type THHW
C. Type XHHW
D. Type THW

ANSWER ______________

PROCEDURE TO ESTABLISH ANSWER

The question is about conductors in a wet location.

1. In Index, find "Conductors," under which, find "Wet locations, 225.4, 310.10(C)."

2. Section 310.10(C) and Table 310.4(A) recognizes Types THHW, THW, and XHHW conductors for wet locations but not Type SA (Silicone) , is only acceptable in dry and damp locations. The "W" denotes the conductor suitability for use in wet locations. Most THHN conductors are dual rated as THHN/THWN, where THHN is used in dry and damp locations in the 90°C column and THWN is for wet locations in the 75°C column.

3. The correct answer is A.

30 What is the minimum size Type THW conductors permitted for a 7.5 m (25 ft) feeder tap from a 500-kcmil feeder protected by a 400-ampere circuit breaker? There is no overcurrent protection at the point of the tap.

A. 2 AWG
B. 1 AWG
C. 1/0 AWG
D. 2/0 AWG

ANSWER _______________

PROCEDURE TO ESTABLISH ANSWER

The question is about size of conductors tapped from a feeder.

1. In Index, find "Feeders" under which, find "Taps, 240.21(B), 240.92(B), 430.28, etc.."

2. Section 240.21(B)(1) refers to 3 m (10 ft) taps which is not applicable. Find Section 240.21(B)(2) which applies to 7.5 m (25 ft) taps and provides that the ampacity of the tap conductors is to be not less than 1/3 of the rating of the overcurrent device protecting the feeder conductors.

3. The tap conductor must have a minimum ampacity of 400 Amperes x .333 = 133 amperes.

4. Referring to Table 310.16 for allowable ampacities of conductors, find that a 1 AWG Type THW copper conductor has an ampacity of 130 amperes, while a 1/0 AWG is rated at 150 amperes.

5. The correct answer is C.

   Note: Although the question did not specify the conductor material, Section 110.5 states that where the material is not specified, to assume it is copper.

31

Six 6 AWG THW copper conductors are installed in the same raceway with six 6 AWG Type XHHW-2 copper conductors. All conductors are current carrying. The permitted ampacity for the Type XHHW-2 conductors is:

A. 25.0 amperes
B. 32.5 amperes
C. 38.5 amperes
D. 45.5 amperes

ANSWER ______________

PROCEDURE TO ESTABLISH ANSWER

This question is about conductor ampacities with conductors of two different temperature ratings in the same raceway.

1. In Index, find "Conductors," under which, find "Ampacities, 310.14 through 310.21."

2. Section 310.15(B) refers to Tables310.16 through 310.21.

3. Section 310.15(C)(1) and Table 310.15(C)(1) requires an adjustment factor to be applied for more than three current-carrying conductors in a raceway or cable.

4. The ampacity for ten through twenty conductors is a multiplier of 50 percent of the conductor ampacity. Six conductors plus six conductors equal twelve conductors in the raceway.

5. In Index, find "Conductors," under which, find "Temperature, limitations of, 310.14(A)(3) , 338.10(B)(3)."

6. Section 310.14(A)(3) limits ampacity of 6 AWG Type XHHW-2 conductors to the same ampacity as 6 AWG Type THW which is listed as 65 amperes in Table 310.16. The -2 after XHHW (XHHW-2) means that the conductor is rated for 90°C applications, whereas XHHW is rated for 75°C applications. See Table 310.4(A) for the operating temperatures of conductors.

7. Going back to 310.15(C)(1) and adjusting the ampacity because of more than three current-carrying conductors in the raceway (derating):

   65 amperes x 50 percent adjustment (12 conductors in the raceway) = 32.5 amperes.

8. The correct answer is B.

32

Which one of the following statements about 2 AWG, Type THHN copper conductors is true?

A. It has a maximum operating temperature of 90°C.
B. It has nylon insulation only.
C. It has an area of .067 square inch.
D. It has a dc resistance of .266 ohms per 300 m (1000 ft).

ANSWER ______________

PROCEDURE TO ESTABLISH ANSWER

The question is about properties of conductors.

1. In Index, find "Conductors," under which, find "Temperature, limitations of, 310.14(A)(3), 338.10(B)(3)."
2. In the Informational Note to 310.14(A)(3), find the reference to Table 310.4(A) .
3. In Table 310.4(A), find Type THHN, which notes a maximum operating temperature 90°C. Answer A is true.
4. First letter T = Thermoplastic. The conductor has thermoplastic insulation. Answer B is not true [see Table 310.4(A)].
5. In Index, find "Conductors," under which, find "Dimensions of, Chap. 9, Table 4." Note: "Table 4" in Index reference should be "Table 5."
6. In Chapter 9, Table 5, find Type THHN and 2 AWG which has an area of 0.1158 in^2. Answer C is not true.
7. In Index, find "Conductors," under which, find "Properties of, Chap. 9, Table 8."
8. In Table 8 under copper, uncoated ohm/kFT find for 2 AWG = 0.194 ohms per 1000 FT. Answer D is not true.
9. The correct answer is A.

33 A Type THW copper conductor with 75°C maximum operating temperature rating is installed in a location with an ambient temperature of 100°F. The correction factor for determining ampacity that must be applied to this conductor is _______?

A. 82 percent
B. 88 percent
C. 90 percent
D. 91 percent

ANSWER _______________

PROCEDURE TO ESTABLISH ANSWER

The question is about temperature ratings of conductor insulations and ambient temperatures.

1. In Index, find "Conductors," under which, find "Temperature, limitations of, 310.14(A)(3), 338.10(B)(3)."
2. Section 310.14(A)(3) Informational Note concerning temperature rating of conductors refers you to the ampacity tables of Article 310 and the correction factors in 310.15(B).
3. In Table 310.15(B)(1), "Ambient Correction Factors Based on 30°C (86°F)." Note in right-hand column degrees in Fahrenheit, 96–104 degrees, and under the copper 75°C vertical column for Type THW copper find .88 multiplier as the correction factor.
4. The correct answer is B.

34

Several Type THW, TW and XHHW conductors are installed in the same conduit. Which one of the following statements is true?

A. The ampacities of the Type THW conductors are figured from the 75°C column of Table 310.16.
B. The ampacities of the Type TW conductors are figured from the 90°C column of Table 310.16
C. The ampacities of the Type XHHW conductors are figured from the 90°C column of Table 310.16.
D. The ampacities of the Type XHHW conductors are figured from the 60°C column of Table 310.16.

ANSWER _______________

PROCEDURE TO ESTABLISH ANSWER

The question is about the temperature limitations on conductors.

1. In Index, find "Conductors," under which, find "Temperature, limitations of, 310.14(A)(3), 338.10(B)(3)."
2. Section 310.14(A)(3) indicates that no conductor shall be used in such a manner that its operating temperature exceeds that designated for the type of insulated conductor involved. In no case shall conductors be associated together in such a way that the limiting temperature of any conductor be exceeded.
3. Since there are 60°C, 75°C and 90°C conductors in the raceway, the 60°C conductor rating must then be used for ampacity determination for all of the conductors including the Types THW, TW and XHHW. This is an important Code concept and requirement to remember.
4. The correct answer is D.

35

In paralleling conductors of size 1/0 AWG or larger, which one of the following conditions is NOT true?

A. They shall be of the same length and circular mil area.
B. They shall be of the same conductor material and insulation.
C. They shall be of the sunlight resistant type.
D. They shall be terminated in the same manner.

ANSWER _______________

PROCEDURE TO ESTABLISH ANSWER

The question is about paralleled conductors.

1. In Index, find "Conductors," under which, find "Paralleled, 250.30(A)(3), 250.102(C), 250.122(F), 250.186(A)(2), 250.186(B)(2), 300.3(B)(1), 300.5(I), Exc. 1, 300.20, 310.10(G) , etc."
2. Sections 310.10(G)(1) through (6) contain the requirements for paralleling conductors of 1/0 AWG and larger, and there are no requirements for sunlight resistance specified. Answer C is not true. Answers A, B, and D are true.
3. Although sunlight resistance was not specified in the question, the requirements are found in Section 310.10(D). The requirements only apply when conductors are exposed to sunlight.

4. The correct answer is C.

36

Branch circuit conductors, installed under raised floors in data processing rooms and that qualify as information technology equipment (ITE) rooms that supply receptacles, are NOT permitted to be installed in:

A. Flexible metal conduit
B. Type NM cable
C. Type AC cable
D. Electrical nonmetallic tubing

ANSWER ____________

PROCEDURE TO ESTABLISH ANSWER

The question is about wiring methods for data processing systems.

1. In Index, find "Data processing systems. see Information technology equipment." In Index, find "Information technology equipment," under which, find "Supply circuits and interconnecting cables, 645.5."

2. Section 645.5(E)(1)(b) specifies the acceptable wiring methods, in addition to those permitted in 300.22(C), which do not include Type NM cable as a wiring method for under raised floors.

3. The correct answer is B.

37

An 800-ampere overcurrent device protects a feeder consisting of two paralleled 500-kcmil conductors per phase in two rigid nonmetallic conduits. Which one of the following statements regarding equipment grounding conductors installed in the conduit(s) for this feeder is true?

A. A single 3 AWG equipment grounding conductor may be installed in one of the conduits.
B. A single 1/0 AWG equipment grounding conductor may be installed in one of the conduits.
C. A 3 AWG copper equipment grounding conductor may be installed in each one of the conduits.
D. A 1/0 AWG copper equipment grounding conductor is required in each one of the conduits.

ANSWER ____________

PROCEDURE TO ESTABLISH ANSWER

The question is about size and installation of equipment grounding conductors for a circuit consisting of paralleled phase conductors.

1. In Index, find "Equipment grounding conductors," under which, find "Sizing, 250.122, 250.190(C)(3)."

2. Section 250.122(F)(1) requires that where conductors are run in parallel in multiple raceways, the equipment grounding conductor, where used, shall be run in parallel. In addition, each parallel equipment grounding conductor must be sized on the basis of the ampere rating of the overcurrent device protecting the paralleled circuit conductors in accordance with Table 250.122.

3. In Table 250.122 for an 800-ampere overcurrent device, the minimum size of the equipment grounding conductor is 1/0 copper, which must be installed in each of the two conduits.

4. The correct answer is D.

38

What minimum size electrical metallic tubing is required for the following nine individual copper conductors: four 12 AWG Type THWN, two 8 AWG Type RHW, and three 6 AWG Type RHW?

A. Metric Designator 27 (Trade Size 1)
B. Metric Designator 35 (Trade Size 1¼)
C. Metric Designator 41 (Trade Size 1½)
D. Metric Designator 53 (Trade Size 2)

ANSWER ______________

PROCEDURE TO ESTABLISH ANSWER

The question is about conductor fill in electrical metallic tubing.

1. In Index, find "Conductor fill," under which, find "Electrical metallic tubing, 358.22."

2. Section 358.22 refers to Table 1, Chapter 9, which permits a 40 percent raceway fill for more than two conductors.

3. Note 6 to Table 1 provides that for combinations of conductors of different size, use the actual size of the conductor or Tables 5 and 5A in Chapter 9 for dimensions of conductors and Table 4 for applicable conduit or tubing dimensions. Since no actual sizes were given in the question, we will use the Tables.

4. Referring to Table 5 for conductor area,

   12 AWG Type THWN area of 0.0133 in^2 x 4 conductors = 0.0532 in^2

   8 AWG Type RHW area of 0.0835 in^2 x 2 conductors = 0.1670 in^2

   6 AWG Type RHW area of 0.1041 in^2 x 3 conductors = 0.3123 in^2

   Total area for 9 conductors (0.0532 in^2 + 0.1670 in^2 + 0.3123 in^2) = 0.5325 sq. in.

5. Referring to Table 4 for "Electrical Metallic Tubing" under the "Over 2 Wires 40%" column:

   A metric designator 27 (Trade Size 1) EMT is permitted to have a 40% conductor fill of 0.346 sq. in.

   A metric designator 35 (Trade Size 1¼) EMT is allowed to have a 40% conductor fill of 0.598 sq. in.

6. The correct answer is B.

39 A 300-kcmil, concentric, stranded conductor has how many strands?

A. 37
B. 61
C. 91
D. 127

ANSWER _______________

PROCEDURE TO ESTABLISH ANSWER

The question is about properties of conductors.

1. In Index, find "Conductors" under which, find "Properties of, Chap. 9, Table 8."

2. Find Chapter 9, Table 8 and in the "Area Cir. Mills" column find 300 kcmil. Read across to "Stranding Quantity" column and find 37.

3. The correct answer is A.

40 A continuous duty rated motor, used for intermittent service to supply a single passenger elevator, requires branch-circuit supply conductors with an ampacity not less than what percentage of the motor nameplate current rating?

A. 85
B. 100
C. 125
D. 140

ANSWER _______________

PROCEDURE TO ESTABLISH ANSWER

The question is about the size of branch-circuit conductors for a single motor used for intermittent service.

1. In Index, find "Motors" under which, find "Conductors, 430–II" under which find "Single motor, 430.22."

2. Scan Article 430 Part II, and find Section 430.22(A), which requires branch-circuit conductors supplying a single motor to have an ampacity not less than 125 percent of the motor full-load current (FLC) rating.

3. Section 430.22(E), "Other Than Continuous Duty," however, requires that conductors for motors used for intermittent duty shall have an ampacity not less than the percentage shown in Table 430.22(E).

4. Table 430.22(E) requires that for a continuous rated motor used for intermittent duty for passenger elevators, the ampacity of the branch-circuit conductors must be not less than 140 percent of the motor nameplate current rating.

5. The correct answer is D.

Note: Section 620.13, "Feeder and Branch-Circuit Conductors" in Article 620 covering elevators, etc., requires conductors to have an ampacity in accordance with 620.13(A)–(D). Section 620.13(A) requires conductors supplying a single motor to have an ampacity not less than the percentage of the motor nameplate current determined from 430.22(A) and (E).

41 3/0 AWG Type THHW copper conductors are used for the branch circuit from a circuit breaker to a motor controller. The terminals on the breaker are rated for 75°C while the terminations on the controller are rated at 60°C. What is the maximum ampacity permitted for the 3/0 AWG conductors?

A. 145 amperes
B. 165 amperes
C. 200 amperes
D. 225 amperes

ANSWER ______________

PROCEDURE TO ESTABLISH ANSWER

The question is about ampacity of conductors based on conductor terminal connections.

1. In Index, find "Terminals," under which, find "Connections to, 110.14, etc."

2. Scan Section 110.14 and find "(C) Temperature Limitations."

3. Section 110.14(C) provides that the temperature rating associated with the ampacity of a conductor shall be selected and coordinated so as not to exceed the lowest temperature rating of any connected termination.

4. Since the 3/0 AWG conductor is connected to a terminal rated at 60°C, the ampacity of the conductor must be based on the Table 310.16 60°C ampacity column for the 3/0 AWG conductor, or in this case, 165 amperes.

5. The correct answer is B.

42 A separately derived system originating in a step-down transformer supplies a 3-phase, 208Y/120V, panelboard. Both the panelboard and the transformer are located inside the building. The grounding electrode conductor for this separately derived system shall be connected to the grounded conductor of the derived system at which one of the following locations?

A. The secondary grounded conductor (neutral) within the transformer enclosure, with the system bonding jumper located at the panelboard.
B. The grounded (neutral) conductor within the panelboard enclosure, with the system bonding jumper located at the transformer.
C. The grounded conductor (neutral) within a junction box installed between the transformer and the panelboard with the system bonding jumper installed in the transformer.
D. At the same point on the separately derived system where the system bonding jumper is installed.

ANSWER ______________

PROCEDURE TO ESTABLISH ANSWER

The question is about the grounding of separately derived systems.

1. In Index, find "Grounding," under which, find "Separately derived systems, 250.21(A), 250.30."

2. Scan Section 250.30(A)(5), which specifies the grounding electrode conductor is to be connected to the grounded conductor of the derived system at the same point where the system bonding jumper is connected to the separately derived system. Since the system bonding jumper is not located at the point where the grounding electrode conductor connects to the system, Answers A, B and C are not true.

3. The correct answer is D.

   Note: Section 250.30(A)(1) permits the system bonding jumper to be placed at any single point on the separately derived system from the source to the first disconnecting means, in accordance with 250.30(A)(1)(a) or (b).

43

A grounding electrode conductor extending from the service equipment to a grounding electrode is judged to be free from physical damage. What is the minimum size grounding electrode conductor that may be installed without having to closely follow the surface of the building finish?

A. 8 AWG
B. 6 AWG
C. 4 AWG
D. 2 AWG

ANSWER ______________

PROCEDURE TO ESTABLISH ANSWER

The question is about the installation of a grounding electrode conductor subject to physical damage.

1. In Index, find "Grounding electrode conductors," under which, find "Installation, 250.64."

2. Section 250.64(B) requires that:

   — A 6 AWG free from exposure to physical damage is permitted to be rigidly fastened to building surface, etc.

   — Grounding electrode conductors smaller than 6 AWG shall be installed in rigid metal conduit, intermediate metal conduit, etc.

3. Since the conductor is not exposed to physical damage, a 6 AWG conductor could be used but it would be required to closely follow the surface of the building construction. However, a 4 AWG conductor would not be required to closely follow the surface of the building construction.

4. The correct answer is C.

44

What maximum distance is permitted for spacing supports for 350-kcmil copper conductors installed vertically in raceways?

A. 12 m (40 ft)
B. 18 m (60 ft)
C. 28 m (95 ft)
D. 36 m (120 ft)

ANSWER ______________

PROCEDURE TO ESTABLISH ANSWER

The question is about the support of conductors in vertical raceways.

1. In Index, find "Conductors," under which, find "Support of, in vertical raceways, 300.19."

2. Table 300.19(A) requires support of the 350 kcmil copper conductors at a minimum of 18 m (60 ft).

3. The correct answer is B.

45

Which one of the following statements about parallel equipment grounding conductors installed with parallel circuit conductors in separate raceways is NOT true?

A. The conductors are required to be size 1/0 AWG or larger.
B. The conductors are required to be of the same length.
C. The conductors shall be of the same size in circular mil area.
D. The raceways shall have the same electrical characteristics.

ANSWER ______________

PROCEDURE TO ESTABLISH ANSWER

The question is about parallel equipment grounding conductors.

1. In Index, find "Paralleled conductors. see Conductors, Parallel."

2. In Index, find "Conductors," under which, find "Paralleled, 250.30(A)(3), 250.102(C), 250.122(F), 300.5(I), 300.20, 310.10(G) , etc."

3. Scan Section 310.10(G)(1) and Informational Note that, where equipment grounding conductors are used with conductors in parallel, equipment grounding conductors shall comply with the requirements for paralleled conductors in Section 310.10(G)(1) through (G)(6) , except that the equipment grounding conductors shall be sized in accordance with Section 250.122.

4. Section 310.10(G)(1) and (G)(3) require the parallel conductors to be of the same length and size in circular mil area and 310.10(G)(3) requires the parallel conductors to have the same electrical characteristics where run in separate raceways or cables. The size of the equipment grounding conductors, however, is based on Table 250.122, and the 1/0 AWG minimum does not apply.

5. The correct answer is A.

46 A 120/240-volt, 3-wire feeder, supplying 120- and 240-volt single-phase equipment with all terminations rated at 75°C, has a calculated load of 150 amperes for both ungrounded conductors and 110 amperes for the neutral conductor. The feeder consists of three copper conductors installed in a raceway indoors. Which one of the following combination of conductors is permitted?

A. Two 1 AWG Type THHN ungrounded (phase) conductors with a 2 AWG Type THW grounded conductors
B. Two 1 AWG Type THHN ungrounded (phase) conductors with a 1 AWG type THHN grounded conductors
C. Two 1/0 AWG Type THW ungrounded (phase) conductors with a 3 AWG Type THHN grounded conductors
D. Two 1/0 AWG Type THW ungrounded (phase) conductors with a 2 AWG Type THW grounded conductors

ANSWER ______________

PROCEDURE TO ESTABLISH ANSWER

The question is about the operating temperature of conductors. Note that some of the answers to the question may be addressing two different temperature ratings of conductor insulation, namely, Type THW at 75°C and Type THHN at 90°C.

1. In Index, find "Conductors," under which, find "Temperature, limitations of, 310.14(A)(3), etc."

2. Section 310.14(A)(3) does not permit any conductor to be used in such a manner that its operating temperature will exceed that designated for the type of insulated conductor involved. Furthermore, this section provides that in no case shall conductors be associated together in such a way that the limiting temperature of any conductor is exceeded.

3. Where conductors of different temperature rating are contained within the same raceway, the ampacity applied to any conductor cannot exceed that required for a conductor having a temperature rating, based on the lowest temperature, of any conductor in the raceway. For example, if 90°C conductors are installed within the same raceway with 75°C conductors, the ampacity of all conductors must be based on the 75°C rating of that conductor.

4. Furthermore, since the terminations on the equipment are rated for 75°C, no conductor may be used at an ampacity greater than its 75°C rating (see Section 110.14).

5. Review the respective conductor ampacities in Table 310.16 and note the following:
   (a) The ampacities of both of the conductors in answers A and C must be based on 75°C because of the Type THW 75°C conductors.

   (b) The ampacities of the conductors in answer B must be based on 75°C because of the 75°C terminations.

   (c) The ampacities of the 1/0 AWG and 2 AWG Type THW in answer D are 150 and 115 amperes, respectively, and are acceptable for the stated loads. The ampacity of the ungrounded conductors in answers A and B and the grounded conductor in answer C is too small.

6. The correct answer is D.

47 Which one of the following statements about 12 AWG Type THW copper conductors installed in a raceway is NOT true?

A. Their ampacity must be adjusted where subjected to ambient temperatures higher than 30°C.
B. They are limited to a 20-ampere maximum overcurrent device in all cases.
C. They must be adjusted (derated) where more than three current-carrying conductors are in the same raceway.
D. They have an ampacity of 25 amperes.

ANSWER _______________

PROCEDURE TO ESTABLISH ANSWER

The question is about ampacities and overcurrent protection for conductors.

1. In Index, find "Conductors," under which, find "Copper," under which, find, "Ampacities, 310.14 through 310.21 etc."

2. Table 310.16 has an asterisk at each of the ampacity ratings for the 12 AWG conductors that refers to the asterisk note at the bottom of the table with reference to Section 240.4(D).

3. Section 240.4(D) states that "unless specifically permitted in 240.4(E) or (G), the overcurrent protection shall not exceed that required by (D)(1) through (D)(7)..." The *Code* permits 12 AWG conductors to be protected with overcurrent devices rated greater than 20 amperes, such as for motor circuits in Article 430, [see Table 240.4(G)]. Answer B is not true.

4. Review Table 310.16 and Table 310.15(C)(1); they confirm that answers A, C, and D are true.

5. The correct answer is B.

48 A bare conductor is to be installed in the same raceway with insulated conductors. All conductors are of the same material and AWG size. Which one of the following statements is true?

A. The ampacity of the bare conductor can be increased 10 percent.
B. The ampacity of the insulated conductors must be decreased 10 percent.
C. The ampacity of the bare conductors is limited to that permitted for the lowest temperature rated insulated conductors.
D. The ampacity of the bare conductor is limited to the 60°C rating of the insulated conductor.

ANSWER _______________

PROCEDURE TO ESTABLISH ANSWER

The question is about the ampacity of bare conductors.

1. In Index, find "Conductors," under which, find "Ampacities of, 310.14 through 310.21, etc."

2. Section 310.15(A) refers to the Ampacity Tables 310.16 through Table 310.21, as modified by 310.15(A) through (F) and 310.12.

3. Scan to Section 310.15(D) and find the requirements applicable to the question.

4. Section 310.15(D) states that where bare conductors are used with insulated conductors the temperature shall be equal to the lowest temperature rating of the insulated conductors.

5. The correct answer is C.

49 Where there are three 3-wire 115/230-volt, multiwire branch circuits in a conduit serving multi-outlet cord- and plug-connected loads and the conductors are 10 AWG Type THW copper, what is the maximum rating of the overcurrent protection permitted for these circuits?

A. 20 amperes
B. 25 amperes
C. 30 amperes
D. 35 amperes

ANSWER _____________

PROCEDURE TO ESTABLISH ANSWER

The question is about the rating of the overcurrent protection permitted where there are more than three branch circuit conductors in a raceway. In order to arrive at this rating, it is necessary to find the ampacity rating of the conductor.

1. In Index, find "Conductors," under which, find "Ampacities of, 310.14 through 310.21 etc."

2. Section 310.15(A) refers to Ampacity Table 310.16 through Table 310.21, as modified by 310.15(A) through (F) and 310.12.

3. Review Section 310.15(E)(1) states that neutral conductors carrying only the normal unbalanced current need not be counted when applying the provisions of Section310.15(C)(1). Also, Section 310.15(C)(1) requires that the conductor ampacities given in Table 310.16 be reduced according to Table 310.15(C)(1).

4. Since there are three circuits involved, there are six 10 AWG Type THW ungrounded current-carrying conductors; and Section 310.15(C)(1) and Table 310.15(C)(1) requires an 80 percent (multiplier) adjustment factor be applied. The conductors have an ampacity of 35 amperes (given in Table 310.16 reduced to 80% by using an 80% multiplier) (35 A x .8 = 28 amperes).

5. In Index, find "Conductors," under which, find "Overcurrent protection, 240.4."

6. Section 240.4(B) permits the next higher rated overcurrent device where it does not correspond to the reduced ampacity rating of the conductor, only where it is not part of a multioutlet branch circuit supplying cord- and plug-connected portable loads. Standard overcurrent device ratings are provided in 240.6 which notes the next lower standard size is 25 amperes. However, 210.18 requires the rating of branch circuits, other than individual branch circuits, to be 15, 20, 30, 40 or 50 amperes. These values are based on receptacle device ratings (based on the overcurrent protective device) and we don't have 25-ampere rated receptacle devices so the next lowest value is 20-ampere.

7. Therefore, in this case, the overcurrent device rating is required to be not greater than 20 amperes.

8. The correct answer is A.

50

A conduit contains eight conductors, supplying two 480Y/277-volt, 3-phase, 4-wire wye-connected lighting branch circuits, serving only fluorescent luminaires. The eight conductors are six ungrounded conductors and two grounded conductors. The ampacity of the conductors must be adjusted to what percentage of the value listed in Table 310.16?

A. 50 percent
B. 60 percent
C. 70 percent
D. 80 percent

ANSWER ______________

PROCEDURE TO ESTABLISH ANSWER

The question is about the derating of conductors and counting of the neutral conductors as current-carrying conductors.

1. In Index, find "Neutral," under which, find "Conductor, 310.15(E), etc."

2. Scan Article 310 to find Section 310.15(E). Find 310.15(E)(3), which is applicable to the question.

3. Section 310.15(E)(3) requires neutrals of 4-wire, 3-phase wye circuits to be counted as current-carrying conductors where serving nonlinear loads, such as electric-discharge lighting.

4. Since there are eight current-carrying conductors, Section 310.15(C)(1) and Table 310.15(C)(1) requires that all conductors be adjusted by a multiplier of 70 percent of the ampacity given in the ampacity tables.

5. The correct answer is C.

51

On a 3-phase, 4-wire wye-connected feeder that supplies primarily data processing equipment, the neutral conductor size may be reduced to what percentage of the size of the phase conductors?

A. 60 percent
B. 70 percent
C. 80 percent
D. No reduction is permitted.

ANSWER ______________

PROCEDURE TO ESTABLISH ANSWER

The question is about the sizing of the neutral conductor for circuits supplying nonlinear loads.

1. In Index, find "Neutral," under which, find "Feeder and service load, 220.61."

2. Neither Section 220.61(C)(2) nor 310.15(E)(3) allow reduction of neutral size for nonlinear loads such as data processing equipment.

3. The correct answer is D.

52 Which one of the following statements about an equipment grounding conductor is true?

A. The conductor is not counted when determining the ampacities of more than three current-carrying conductors in a raceway or cable.
B. The conductor is counted when determining the ampacities of more than three current-carrying conductors installed in a raceway.
C. The conductor is not counted when determining conduit fill in a nonmetallic conduit.
D. The conductor is not counted when determining conduit fill in flexible metal conduit.

ANSWER ______________

PROCEDURE TO ESTABLISH ANSWER

The question is about conductor ampacity and conduit fill when utilizing an equipment grounding conductor.

1. In Index, find "Conduit fill," under which, find "Equipment grounding conductors, Chap. 9, Note 3 to Table 1."

2. Note 3 requires the equipment grounding conductor to be included when calculating conduit or tubing fill. The type of raceway is not a factor. Answers C and D are not true.

3. In Index, find "Conductors," under which, find "Ampacities of, 310.14 through 310.21, etc."

4. Section 310.15(A) refers to the Ampacity Table 310.16 through Table 310.21, as modified by 310.15(A) through (F) and 310.12.

5. Scan the and find 310.15(F) which states that equipment grounding conductors shall not be counted when applying the provisions of 310.15(C)(1) and Table 310.15(C)(1) for more than three current-carrying conductors in a raceway or cable. Answer B is incorrect.

6. The correct answer is A.

53 Open wiring on insulators is required to be separated from metal conduit, piping or other conducting material by not less than what distance?

A. 25 mm (1 in.)
B. 50 mm (2 in.)
C. 75 mm (3 in.)
D. 100 mm (4 in.)

ANSWER ______________

PROCEDURE TO ESTABLISH ANSWER

The question is about the separation of open wiring on insulators from conducting material.

1. In Index, find "Open wiring on insulators," under which, find "Clearances, 398.19."

2. Section 398.19 requires open wiring on insulators to be separated 50 mm (2 in.) from metal raceways, piping, or other conducting material.

3. The correct answer is B.

54 The ampacity of each 4/0 AWG Type THW-2 copper conductor installed in EMT where there are six current-carrying conductors is:

A. 195 amperes
B. 208 amperes
C. 230 amperes
D. 260 amperes

ANSWER ______________

PROCEDURE TO ESTABLISH ANSWER

The question is about the ampacity of Type THW-2 insulated conductors. It is important to recognize that the insulation in the question is Type THW-2, not Type THW. Those who are not familiar with electrical installations might assume that Type THW-2 insulation is suitable for 75°C, rather than 90°C. This demonstrates the importance of reading examination questions carefully. THW-2 is rated for 90°C ampacity.

1. In Index, find "Conductors," under which, find "Ampacities of, 310.14 through 310.21 etc."

2. In Section 310.14(A), find that ampacities for conductors can be determined by tables or under engineering supervision. Since the question does not refer to engineering supervision, assume that the ampacity will be determined from the tables.

3. Look in 310.15(A) that states the "Ampacities for conductors rated 0 volts to 2000 volts shall be as specified in the Ampacity Table 310.16 through Table 310.21, as modified by 310.15(A) through (F) and 310.12.

4. Scan and find the "Adjustment Factors" in 310.15(C)(1) and Table 310.15(C)(1) that are applied where there are more than three current-carrying conductors in a raceway or cable.

5. Find 4/0 AWG Type THW-2 conductors in Table 310.16 where the allowable ampacity is 260 amperes in the 90°C column.

6. Table 310.15(C)(1) requires that an adjustment factor (multiplier) of 80 percent be applied to each conductor where the number of current-carrying conductors in a raceway is between 4 and 6.

7. 260 amperes x 0.8 (80 percent) = 208 amperes.

8. The correct answer is B.

55

Where more than one nominal voltage electrical system exists in a building, which statement is NOT true for identification of feeder conductors for different systems?

A. Black, red, blue colored marking is required for conductors of 208Y/120-volt systems; and brown, orange, yellow is required for 480Y/277-volt systems.
B. Each conductor of a feeder, where accessible, shall be identified by phase or line and system at all termination, connection, and splice points.
C. The identification scheme for the ungrounded conductors must be permanently posted at each feeder panelboard or similar distribution equipment in the building.
D. The means of identification may be by separate color coding, marking tape, tagging, or other approved means.

ANSWER ____________

PROCEDURE TO ESTABLISH ANSWER

The question relates to the identification of feeder conductors.

1. In Index, find "Conductors," under which, find "Feeder, Art. 215"

2. Scan Article 215, "Feeders," and find Section 215.12, "Identification of Feeders." Find Section 215.12(C), "Ungrounded Conductors," and answers B, C, and D are found to be included in the section. Answer A is not true.

3. The correct answer is A.

56

Which is NOT true of temperature limitations of conductors?

A. In no way may a conductor be used in excess of its designed insulation temperature rating
B. The rate at which heat dissipates into the ambient medium
C. Adjacent cables or conductors placed closely together have the dual effect of raising the ambient temperature and impeding heat dissipation
D. Harmonic currents do not affect the ampacity rating of a conductor

ANSWER ____________

PROCEDURE TO ESTABLISH ANSWER

This question addresses the question of what can impact the temperature limitations of conductors.

1. In Index, find "Temperature Limitations," under which, find "Conductors, 310.14(A)(3), 338.10(B)(3)." Scanning through 310.14(A)(3), and Informational Note No. 1, we see that answers A, B, and C are true. Informational Note No. 1, (2) proves that Answer D is not true, and that harmonic current does affect temperature in conductors.

2. The correct answer is D.

57 Which of the following is NOT true of ampacity adjustments for conductors?

A. Adjustment factors (derating) do not apply to conductors in nipples 24 inches long.
B. Adjustment factors (derating) do not apply to underground conductors entering or leaving an outdoor trench if those conductors have physical protection in PVC conduit and do not have more than four conductors in the raceway.
C. A 3-phase feeder installed with two conductors in parallel per phase does not require ampacity adjustment.
D. Each current-carrying conductor of a paralleled set of conductors shall be counted as a current-carrying conductor.

ANSWER ______________

PROCEDURE TO ESTABLISH ANSWER

This question is about when ampacity adjustment (derating) is required.

1. In Index, find "Conductors," under which, find "Ampacities of, 310.14 through 310.21, etc."

2. Scan through 310.15 to find "310.15(C)(1), Adjustment Factors." Subsections 310.15(C)(1)(b) and 310.15(C)(1)(c) provide that answers A and B are true. Answer D is true because each current-carrying conductor of a paralleled set of conductors shall be counted as a current-carrying conductor per 310.15(C)(1) last sentence. It is important to note that where paralleling conductors are installed in the same cable or raceway, ampacity adjustment due to the number of current carrying conductors may be required. Answer C is not true.

3. The correct answer is C.

58 Cables or raceways installed under roof decking shall be installed and supported so the distance from the lowest surface of the roof decking to the top of the cable or raceway is not less than:

A. 32 mm (1¼ in.)
B. 38 mm (1½ in.)
C. 44 mm (1¾ in.)
D. This is not a *Code* requirement.

ANSWER ______________

PROCEDURE TO ESTABLISH ANSWER

The question is about protection from physical damage for cables or raceways under roof decking.

1. In Index, find "Cables," under which, find "Protection against physical damage, 300.4."

2. In Sections 300.4, find 300.4(E), "Cables and Raceways Installed Under Roof Decking."

3. The answer is 38 mm (1½ in.) from the lowest surface of roof decking.

4. The correct answer is B.

Note: The problem often occurs after the cables and raceways are installed and where, at some time much later, the building is re-roofed. Often insulating material and a new roof membrane are added and longer screws are used to secure these materials to the existing corrugated steel roof deck. This is when the cables and raceways may be penetrated by a screw. This also applies to luminaires and boxes.

59

The interior of raceways installed underground shall:

A. be considered a wet location.
B. be free and dry with no moisture in them.
C. be considered a damp location.
D. be considered a dry location.

ANSWER ______________

PROCEDURE TO ESTABLISH ANSWER

The question is about the interior of raceways where installed underground.

1. In Index, find "Underground wiring, 300.5."

2. In Section 300.5, find 300.5(B), which states that the interior of raceways installed underground are considered a wet location.

3. In Index, find "Damp or wet locations," under which, find "Definition, Article 100." In Article 100, find *Locations, Wet* which includes installations underground or in concrete slabs or masonry in direct contact with the earth, etc.

4. The correct answer is A.

CHAPTER 7
SERVICE EQUIPMENT

1 A 3-phase, 3-wire delta-connected service is installed. One of the phase conductors is grounded to a grounding electrode. Which one of the following statements is true?

A. The service overcurrent protective device is required in series with the grounded phase conductor.
B. A feeder and branch-circuit overcurrent protective device is required to be installed in series with the grounded phase conductor.
C. The service disconnecting means and overcurrent protection may consist of a 3-pole disconnect switch with three fuses.
D. The service disconnecting means and overcurrent protection may consist of a 3-pole common trip circuit breaker.

ANSWER ______________

PROCEDURE TO ESTABLISH ANSWER

The question is about overcurrent protection in a grounded conductor.

1. In Index, find "Overcurrent protection," under which, find "Location, 230.91, 230.92, 240.92, 240–II."."

2. Under Article 240, Part I, find Section 240.15(A), which requires that a fuse or an overcurrent trip unit of a circuit breaker shall be connected in series with each ungrounded conductor. Answer A doesn't differentiate between fuses or circuit breakers and 240.22 would not permit fuses. Answer B is not correct since the question is dealing with service overcurrent protection. Answer C is incorrect since the device must open all of the conductors simultaneously and fuses won't satisfy this requirement. A three-pole common trip in Answer D will open all of the conductors simultaneously and fuses will satisfy this requirement

3. Section 240.22 in Part II requires that no overcurrent device shall be connected in series with any conductor that is intentionally grounded, with two conditions.

4. Section 240.22(1) permits an overcurrent device in the grounded conductor where the overcurrent device opens all conductors, including the grounded conductor, and is designed so that no pole can operate independently.

5. The correct answer is D.

2 A 480Y/277-volt service is grounded at the utility transformer and enters a building. Under which one of the following conditions is it necessary to bring the grounded conductor into the service equipment?

A. Where only one 480-volt motor is installed
B. Where 3-phase motors and 277-volt lighting is installed
C. Where only 480-volt, 3-phase power is needed
D. All of the above

ANSWER ______________

PROCEDURE TO ESTABLISH ANSWER

The question is about bringing a grounded conductor into the service equipment.

1. In Index, find "Grounding," under which, find "AC systems, 250.20, 250.24 through 250.26."

2. Section 250.24(C) requires that, where an ac system operating at less than 1000 volts is grounded at any point, the grounded conductor shall be run to each service disconnecting means and shall be bonded to each disconnecting means enclosure. The type of load is not a factor. The grounded conductor (neutral) conductor must be run to the service disconnecting means to provide a low impedance path back to the transformer for clearing ground faults.

3. The correct answer is D. Note: See 250.186 for over 1000 volts.

3 A 208Y/120-volt, 3-phase, 4-wire service consists of two 500 kcmil AL conductors paralleled per phase. What is the minimum size copper grounding electrode conductor that could be run to a water pipe grounding electrode for this service?

A. 2 AWG
B. 1/0 AWG
C. 2/0 AWG
D. 3/0 AWG

ANSWER ______________

PROCEDURE TO ESTABLISH ANSWER

The question is about the minimum size of a grounding electrode conductor.

1. In Index, find "Grounding electrode conductors," under which, find "Sizing, 250.30(A)(6)(a), 250.66, 250.166."

2. Section 250.66 requires that the grounding electrode conductor size be not less than given in Table 250.66, unless you meet any of the conditions in 250.66(A), (B) or (C). However, none of those conditions apply in this case.

3. Table 250.66 relates to the size of largest service-entrance conductor or equivalent area for parallel conductors (which are the sum of the areas of all conductors in parallel per phase). Then read across to the right to the appropriate column (CU or AL), that then specifies the required size grounding electrode conductor.

4. Thus, two 500 kcmil AL x 2 = 1000 kcmil. Reading across from this figure in the AL column, find the required size of copper grounding electrode conductor is 2/0 AWG.

5. The correct answer is C.

4 Which one of the following statements about a service switchboard or panelboard installed indoors is true?

A. The service switchboard or panelboard shall be installed in rooms dedicated exclusively to such equipment.
B. Equipment foreign to the electrical service switchboard or panelboard shall not be permitted to enter or pass through its dedicated room.
C. Sprinkler protection is permitted for the dedicated space.
D. Sprinkler piping may be located directly above and within 1.8 m (6 ft.) of the top of the switchboard or panelboard.

ANSWER ______________

PROCEDURE TO ESTABLISH ANSWER

The question is about the installation of switchboards and panelboards.

1. In Index, find "Switchboards," under which, find "Installation, indoor and outdoor, 110.26."
2. Scan Section 110.26, and find 110.26(E), Dedicated Equipment Space. Section 110.26(E) requires that switchboards and panelboards be located with dedicated spaces. Foreign equipment (to the electrical installation) is not permitted to be installed in the dedicated width and depth of the area directly above the equipment. There is no requirement for a dedicated room; consequently, there is no prohibition against equipment foreign to the electrical equipment entering or passing through such rooms. Answers A and B are not true.
3. A review of 110.26(E)(1)(a) and (c) reveals that sprinkler protection is permitted for the dedicated space, provided the piping is not located in the dedicated space or working clearance space. Answer C is true; Answer D is not true.
4. The correct answer is C.

5 A service panelboard supplied from 3-phase, 4-wire delta-connected system must meet all of the following conditions EXCEPT:

A. A main bonding jumper is required within the panelboard.
B. The high-leg is required to be marked by an outer finish red in color.
C. The phase busbar or conductor having the higher voltage to ground must be marked or identified.
D. The B phase shall be that phase having the higher voltage to ground.

ANSWER ______________

PROCEDURE TO ESTABLISH ANSWER

The question is about busbars and arrangement of conductors in panelboards.

1. In Index, find "Panelboards," under which, find "Support for busbars and conductors, 408.3" and "Service equipment, 408.3(C) and 408.36, Exc. 1 and 3."
2. Section 408.3 applies to the question.

3. Section 408.3(C) requires each service panelboard to be provided with a main bonding jumper. Answer A is true.

4. Section 408.3(E) requires the phase arrangement on 3-phase buses to be A, B, C from front to back, top to bottom, or left to right, as viewed from the front of the switchboard or panelboard. The B phase shall be that phase having the higher voltage to ground on 3-phase, 4-wire, delta-connected systems. Answer D is true.

5. Notice the Informational Note following Section 408.3(E), which refers to Section 110.15 for marking the phase conductors. Answer C is true.

6. Section 110.15 requires the busbar or conductor having the higher voltage to ground to be marked by an outer finish of orange or by other effective means. Answer B is not true.

7. The correct answer is B.

6 Generally, an entrance or egress from the working space not less than 610 mm (24 in.) wide and 2.0 m (6½ ft) high is required at each end of the working space for a 2.5 m (8 ft) wide switchboard with all of the following ratings EXCEPT:

A. 1000 amperes
B. 1200 amperes
C. 1500 amperes
D. 2000 amperes

ANSWER ______________

PROCEDURE TO ESTABLISH ANSWER

The question is about working space about switchboards.

1. In Index, find "Working space," under which, find "About electrical equipment, 110.26, etc."

2. Section 110.26(C)(2) requires equipment rated 1200 amperes or more and over 1.8 m (6 ft) wide to have one entrance to the required working space not less than 610 mm (24 in.) wide and 2.0 m (6½ ft) high at each end of the working space.

   Note: In the case of a large room, or a location that permits a continuous and unobstructed way of egress travel, a single entrance to or egress from the working space is allowed in 110.26(C)(2)(a). 110.26(C)(2)(b) permits a single entrance or egress from the working space if the depth of the working space in front of the equipment is at least twice the minimum depth required in Table 110.26(A)(1), and the distance from the nearest edge of the entrance or egress to the equipment is the minimum distance required by Table 110.26(A)(1) [one working space]. The objective of the *Code* is to provide a safe and clear entrance to readily access and egress the equipment and overcurrent protective devices/disconnects, and for a clear and unobstructed path for persons working on the equipment to exit the space quickly if arcing, explosion, etc., occurs.

3. The correct answer is A.

7 A multifamily dwelling contains ten dwelling units each having a total area of 1000 square feet. A single 120/240-volt service supplies the building ,and a single 120/240-volt feeder supplies each dwelling unit. Each dwelling unit is provided with the following equipment:

1.5 kVA laundry circuit
10 kVA electric range
4 kVA electric water heater (40 gal.)
8 kVA electric central heat
8 kVA electric central A/C
1.2 kVA electric dishwasher
5 kVA electric clothes dryer

Using the optional calculation method, what is the minimum ampacity required for a single service supplying the building described above? There are no house loads.

A. 640 amperes
B. 817 amperes
C. 1000 amperes
D. 1552 amperes

ANSWER ______________

PROCEDURE TO ESTABLISH ANSWER

The question is about calculating minimum service-entrance ampacity.

1. In Index, find "Loads," under which, find "Feeders," under which find "Calculations for, 215.2(A), Art. 220, Annex D."

2. Scan Article 220, and find "Part IV, Optional Feeder and Service Load Calculations," under which find "220.84 and Table 220.84, Optional Calculation — Multifamily Dwelling."

3. The conditions of the question meet the basic requirements of Section 220.84, with the following calculations to determine the ampacity of a single service to the building.

4. Proceed with the calculations.

Lighting and general-use receptacle load
1000 sq. ft. x 3 VA per sq. ft.
= 3000 VA, or 3 kVA per unit

Lighting loads
10 units x 3000 VA = 30 kVA (thousand volt-amperes)

Laundry circuit
10 units x 1500 VA = 15 kVA

Two appliance circuits at 1.5 kVA
10 units x 3000 VA = 30 kVA

Range

10 units x 10,000 VA = 100 kVA

Water heater

10 units x 4000 VA = 40 kVA

Dishwasher

10 units x 1200 VA = 12 kVA

Clothes dryer

10 units x 5000 VA = 50 kVA

Air conditioner*

10 units x 8000 VA = 80 kVA

Total connected load = 357 kVA

5. Apply demand factor of Table 220.84

   A = 357 kVA x 43% based on 8 to 10 units

   A = 153.5 kVA

6. A = 153, 500 VA/240 V

   A = 640 amperes

7. The correct answer is A.

*Use heat or air-conditioning load, but not both [Sections 220.84(C)(5) and 220.60].

Note: Because both of the load amounts are the same, either the heat or the air-conditioning load may be chosen as the larger of the two loads.

8 An apartment building with a 208Y/120-volt, 3-phase, 4-wire service has 30 units each having 1000 square feet with the following load: electric heat (3 kVA), air conditioner (2 kVA), range (10 kVA), water heater (4 kVA), disposer (0.5 kVA), two vent fans (0.25 kVA) each. There are no laundry facilities. Using the optional calculation method, what is the calculated load for the service equipment?

A. 400 amperes
B. 660 amperes
C. 1300 amperes
D. 1660 amperes

ANSWER ______________

PROCEDURE TO ESTABLISH ANSWER

The question is about calculation of service size using the optional calculation method.

1. In Index, find "Loads," under which, find "Feeders, calculations for, 215.2(A), Art. 220, Annex D."

2. In Article 220, Part IV, Optional Feeder and Service Load Calculations, find "220.84 Multifamily Dwelling."

3. The conditions of the question meet the basic requirements of 220.84 with the following calculations to determine the ampacity of a single service to the building.

4. Proceed with the calculations.

Lighting and general-use receptacles	
30 units x (1000 sq. ft. x 3 VA)	= 90 kVA
Appliance circuits	
30 units x (1500 VA x 2 small appliance circuits)	= 90 kVA
Electric heat	
30 units x 3 kVA	= 90 kVA
Range	
30 units x 10 kVA	= 300 kVA
W/H	
30 units x 4 kVA	= 120 kVA
Disposer	
30 units x 0.5 kVA	= 15 kVA
Vent fans	
30 units x (0.25 kVA x 2 fans)	= 15 kVA
Total connected load	= 720 kVA

Note: In this question, the heating load is larger than the air-conditioning load; therefore, the A/C load is not counted in the calculation. It is considered noncoincidental (both systems should not be operating at the same time).

5. Table 220.84 allows a demand factor of 33 percent for 30 units. Therefore:
   720 kVA x 33% = 237.6 kVA demand load.

   Minimum ampacity
   A = 237,600 VA / 208 V / 1.73
   A = 660 amperes

6. The correct answer is B.

9 Which one of the following wiring methods is NOT acceptable for service-entrance conductors over 1000 volts installed aboveground?

A. Intermediate metal conduit
B. Rigid Polyvinyl Chloride conduit (PVC)
C. Open runs of metal-clad cable suitable for the use and purpose
D. Open runs of Type AC cable

ANSWER ______________

PROCEDURE TO ESTABLISH ANSWER

The question is about wiring methods for service-entrance conductors.

1. In Index, find "Service-entrance conductors," under which, find "Over 1000 volts, 230–VIII."

2. Scan Article 230, Part VIII, and find Section 230.202(B), that requires service-entrance conductors to be installed by one of the wiring methods covered in Sections 300.37 for above ground and 300.50 for underground.

3. Section 300.37 recognizes the wiring methods specified in answers A, B, and C, but not Type AC cable.

4. The correct answer is D.

10 For the purpose of Section 230.40, Exception No. 2 only, three sets of underground conductors size 1/0 AWG, running to the same location and connected together at the supply end but not connected together at the load end shall be considered to be which one of the following?

A. A violation
B. One service
C. Three laterals
D. Three services

ANSWER ______________

E. PROCEDURE TO ESTABLISH ANSWER

The question is about the number of services permitted to a building.

1. n Index, find "Services," under which, find "Number, 230.2."

2. Section 230.2 provides that for the purpose of Section 230.40, Exception No. 2 only, underground sets of conductors size 1/0 AWG and larger, running to the same location and connected together at the supply end but not connected together at the load end shall be considered to be one service.

3. The correct answer is B.

11

What is the minimum size of separate sets of underground service conductors that are permitted to be run from a pad mount transformer to six separate service disconnects grouped at one location and considered as a single lateral?

A. 1/0 AWG
B. 3/0 AWG
C. 250 kcmil
D. 500 kcmil

ANSWER ______________

PROCEDURE TO ESTABLISH ANSWER

The question is about the number of services permitted to a building.

1. In Index, find "Services," under which, find "Number, 230.2"

2. Section 230.2 provides that the minimum size of conductors is 1/0.

3. The correct answer is A.

12

A bare copper grounded (neutral) service-entrance conductor is permitted only if the voltage is not over which one of the following?

A. 150 volts to ground
B. 300 volts to ground
C. 600 volts to ground
D. Any voltage in A, B, or C

ANSWER ______________

PROCEDURE TO ESTABLISH ANSWER

The question is about the use of uninsulated grounded (neutral) service-entrance conductors.

1. In Index, find "Service-entrance conductors," under which, find "Insulation, 230.41."

2. Section 230.41 requires service-entrance conductors entering or on the exterior of buildings to be insulated, unless meeting the conditions in the exception.

3. Section 230.41, Exception, items (1) through (5) provide the conditions under which a grounded (neutral) conductor can be uninsulated. Voltage is not a condition listed.

4. The correct answer is D.

13

Which one of the following is not permitted to be installed with service-entrance conductors in a service raceway?

A. Grounding electrode conductors
B. Metering conductors for CT or PT metering
C. Supply-side bonding jumpers
D. Load management control conductors

ANSWER ____________

PROCEDURE TO ESTABLISH ANSWER

The question is about conductors permitted in service raceways.

1. In Index, find "Service raceways," under which, find "Conductors, others permitted in, 230.7"

2. Section 230.7, Exception Nos. 1 and 2 permit grounding conductors, bonding jumpers, and load management control conductors having overcurrent protection in a service raceway.

3. Metering conductors for CT or PT metering are not mentioned in 230.7, Exception Nos. 1 and 2, and, therefore, are not permitted in the service raceway.

4. The correct answer is B.

14

Which one of the following taps is NOT permitted to be connected to the supply side of the service disconnecting means?

A. For load management devices
B. For a standby power system
C. To supply outside area lighting
D. To supply fire pump equipment

ANSWER ____________

PROCEDURE TO ESTABLISH ANSWER

The question is about connecting equipment ahead of the service disconnecting means.

1. In Index, find "Service-entrance equipment, see Service equipment." Under "Service Equipment," find "Disconnecting means, 230–VI," under which, find "Connections, supply side, 230.82."

2. Section 230.82 in Item (5), conductors are permitted to supply load management devices, standby power systems, and fire pump equipment, if provided with service equipment. Outside area lighting is not included in Items (1) through (11), and, therefore, is not permitted.

3. The correct answer is C.

15

The minimum elevation of an unguarded 13,800-volt bus above the working space in a switchboard room is:

A. 2.1 m (7 ft)
B. 2.5 m (8 ft)
C. 2.7 m (9 ft)
D. 2.9 m (9 ft 6 in.)

ANSWER ______________

PROCEDURE TO ESTABLISH ANSWER

The question is about elevation of unguarded live parts over 1000 volts.

1. In Index, find "Working Space," under which, find "Adjacent to live parts (circuits over 1000 volts), 110.32 through 110.34."
2. Scan Section 110.34 and find 110.34(E), which refers to Table 110.34(E).
3. Table 110.34(E) requires 2.9 m (9 ft 6 in.) elevation for voltages between phases of 7501 Volts and 35 kV.
4. The correct answer is D.

16

Which one of the following statements about service disconnects is true?

A. For installations to supply only limited loads of a single branch circuit, the service disconnecting means shall have a rating of not less than 15 amperes.
B. For installations consisting of not more than two 2-wire branch circuits, the service disconnecting means shall have a rating of not less than 60 amperes.
C. For installations to supply a one-family dwelling, the service disconnecting means shall have a rating of not less than 200 amperes.
D. For a service to a small office building consisting of five branch circuits, the service disconnecting means shall have a rating of not less than 100 amperes.

ANSWER ______________

PROCEDURE TO ESTABLISH ANSWER

The question is about the rating of service disconnecting means.

1. In Index, find "Service-entrance equipment, see Service Equipment." Under "Service Equipment," find "Disconnecting means, 230–VI." Under "Disconnecting means, 230–VI," find "Rating, 230.79, 230.80."
2. Scan Section 230.79 and find that 230.79(A) confirms that statement A is true.
3. A review of 230.79 (B), (C), and (D) verifies that statements B, C, and D are not true.
4. The correct answer is A.

17

Where the service disconnecting means consists of more than one switch or circuit breaker, the combined rating of all the switches or circuit breakers used shall not be less than which one of the following?

A. The load to be carried
B. The rating required for a single switch or circuit breaker
C. The rating required for the computed load
D. 125 percent of the computed load

ANSWER ______________

PROCEDURE TO ESTABLISH ANSWER

The question is about the combined rating of service equipment disconnecting means consisting of more than one switch or circuit breaker.

1. In Index, find "Service equipment," under which, find "Disconnecting means, 230–VI," under which, find "Rating 230.79, 230.80."

2. Section 230.79 does not directly address the question so find Section 230.80, Combined Rating of Disconnects.

3. Section 230.80 states that the combined rating of all switches or circuit breakers used shall not be less than the rating required by Section 230.79, would be the rating required for a single service disconnecting means.

4. The correct answer is B.

18

In a multi-occupancy shopping mall, the service overcurrent protection for each occupancy may be located in any of the following locations EXCEPT:

A. Within each occupancy at a readily accessible location
B. In a common readily accessible location outside
C. In an electrical equipment room that is accessible to and under the supervision of the building management
D. In an electrical metering room accessible only to the serving electric utility

ANSWER ______________

PROCEDURE TO ESTABLISH ANSWER

The question is about the location of service overcurrent devices.

1. In Index, find "Overcurrent protection," under which, find "Location, 230.91, 230.92, 240.92, 240–II." Although the question is not specific to overcurrent devices for "branch circuits," it is reasonable to assume there will be branch circuits. In the index, find "Overcurrent devices, 240.24(B)(2).

2. Scan Article 240 Part II and find Section 240.24(A) Accessibility, and Section 240.24(B) Occupancy.

3. Section 240.24(B)(1) permits the service overcurrent devices to be accessible to authorized management personnel only in multiple-occupancy buildings where the electric service and maintenance are provided by the building management and are under continuous building management supervision. This would not include an electric utility company. Answers A, B, and C are true. Answer D is not true.

4. The correct answer is D.

19

Which one of the following statements about ground-fault protection of equipment (GFPE) for service disconnects rated 1000 amperes or more is NOT true?

A. Protection is required for a 277/480-volt impedance grounded system.
B. The maximum setting of the ground-fault protection shall be 1200 amperes.
C. The maximum time delay shall be one second for ground-fault currents greater than 3000 amperes.
D. The provision for ground-fault protection of equipment (GFPE) on service equipment of 1000 amperes does not apply to services for fire pumps.

ANSWER ______________

PROCEDURE TO ESTABLISH ANSWER

The question is about a service ground-fault protection device.

1. In Index, find "Ground-fault protection," under which, find "Service disconnecting means, 230.95."

2. Section 230.95 requires ground-fault protection of equipment for solidly grounded wye electrical services of more than 150 volts to ground, but not exceeding 1000 volts phase-to-phase for each service disconnect rated 1000 amperes or more.

3. Note the definition of *solidly grounded* in Article 100, which is when the grounded conductor is grounded *without* inserting any resistor or impedance device. Since statement A indicates that the system is impedance grounded, statement A is not true.

4. Section 230.95(A) confirms that statements B and C are true.

5. There is no direct language regarding ground-fault protection of equipment on fire pump services to verify Statement D in 230.95. In the Index, find "Fire pumps, Article 695. Scan Article 695 and find "695.6(G), Ground-Fault Protection of Equipment." Section 695.6(G) prohibits ground-fault protection for fire pump power wiring. Answer D is true.

6. The correct answer is A.

20

Which one of the following services for an industrial building requires ground-fault protection of equipment (GFPE)?

A. A 208Y/120-volt, 3-phase, 4-wire solidly grounded wye service with a single main disconnect switch rated at 1000 amperes with 800-ampere fuses
B. A 480Y/277-volt, 3-phase, 4-wire solidly grounded wye service with a single main disconnect switch rated at 1000 amperes with 800-ampere fuses
C. A 480-volt, 3-phase, 3-wire service with a 1000-ampere main circuit breaker
D. A 480Y/277-volt, 3-phase, 4-wire solidly grounded delta service with a single main disconnect switch rated at 1000 amperes with 1000 ampere fuses

ANSWER ______________

PROCEDURE TO ESTABLISH ANSWER

The question is about what services require ground-fault protection and at what rating.

1. In Index, find "Ground-fault protection," under which, find "Service disconnecting means, 230.95."
2. Section 230.95 requires ground-fault protection of equipment for solidly grounded wye electrical services of more than 150 volts to ground, but not exceeding 1000 volts phase-to-phase for each service disconnect rated 1000 amperes or more.
3. Note the definition of solidly grounded found in Article 100, which is when the grounded conductor is grounded without inserting any resistor or impedance device.
4. Section 230.95 confirms the GFP of equipment is required for service disconnecting means rated 1000 amperes or more with the second paragraph of the section clarifying that the rating of the disconnecting means is considered to be the rating of the largest fuse that can be installed.
5. Answer A is not true because the voltage rating is below the requirement for ground-fault protection (GFP). Answer C is a 3-wire delta service, and Answer D is a 4-wire delta service, neither of which requires GFP since these systems are not "wye-type" services. Also note that there is no such thing as a 277/480-volt 3-phase 4-wire delta service. Voltages would be 240/480, with a 416-volt high-leg.
6. The correct answer is B.

21 Service-entrance conductors rated over 1000 volts are considered protected by which one of the following overcurrent protective devices?

A. A fuse rated six times the ampacity of the conductor
B. A fuse rated eight times the ampacity of the conductor
C. A circuit breaker rated six times the ampacity of the conductors
D. A circuit breaker rated eight times the ampacity of the conductors

ANSWER ______________

PROCEDURE TO ESTABLISH ANSWER

The question is about overcurrent protection of service conductors rated over 1000 volts.

1. In Index, find "Service-entrance conductors," under which, find "Over 1000 volts, 230–VIII."
2. Scan Article 230 Part VIII and find "230.208," which provides the requirements of overcurrent protection.
3. Section 230.208 states that a fuse rated in continuous amperes not to exceed three times the ampacity of the conductor, or a circuit breaker with a trip setting not more than six times the ampacity of the conductors, the ampacity of the conductors is considered as providing the required short-circuit protection.
4. The correct answer is C.

22 Which one of the following statements about a totally enclosed switchboard located within an equipment room and provided with a space of 150 mm (6 in.) between the top of the switchboard and ceiling is true?

A. The installation is permitted only if the ceiling is of noncombustible construction.
B. The installation is permitted only if a noncombustible barrier is provided above the switchboard.
C. The installation is acceptable regardless of ceiling construction.
D. The installation is not acceptable unless there is at least 300 mm (12 in.) from top of switchboard to the ceiling.

ANSWER ______________

PROCEDURE TO ESTABLISH ANSWER

The question is about the minimum space required between the ceiling of the room and the top of a totally enclosed switchboard.

1. In Index, find "Switchboards," under which, find "Clearances, 110.26, 408.5, 408.18."

2. Section 408.18(A) states that, for other than a totally enclosed switchboard, a 900 mm (3 ft) space shall be provided between the top of a switchboard and any combustible ceiling, unless a noncombustible shield is installed. There is no specific ceiling requirement for totally enclosed switchboards.

3. The correct answer is C.

23 Which one of the following statements about service-entrance conductors for an overhead system is true?

A. include the service-drop conductors from a utility.
B. are required to be installed in a raceway.
C. only extend from a meter socket to the weatherhead.
D. extend from the terminals of the service equipment to a point usually outside the building where conductors attach to the service drop or overhead service conductors.

ANSWER ______________

PROCEDURE TO ESTABLISH ANSWER

The question is about service-entrance conductors for an overhead system and how they are defined.

1. In Index, find "Overhead service conductors, "Definition, Art. 100–."

2. Scan Article 100, and find "Service-Entrance Conductors, Overhead System," which confirms that statement D is true.

3. The correct answer is D.

24

What is the minimum headroom required for the working space in front of an 800-ampere service-disconnect switch?

A. 1.8 m (6 ft)
B. 2.6 m (6 ft 3 in.)
C. 2.0 m (6 ft 6 in.)
D. 2.5 m (8 ft)

ANSWER _______________

PROCEDURE TO ESTABLISH ANSWER

The question is about the minimum headroom of the working space in front of electrical equipment.

1. In Index, find "Working space," under which, find "About electrical equipment, 110.26, etc."

2. Scan Section 110.26(A)(3), which requires minimum headroom of the working space about service equipment to be 2.0 m (6½ ft). Where the equipment exceeds 2.0 m (6½ ft) in height, the minimum working space height cannot be less than the height of the equipment.

3. The correct answer is C.

25

More than one service is permitted to supply a building or structure under all of the following conditions EXCEPT:

A. Where the capacity requirements is 1600 amperes
B. For emergency, legally required standby, or optional standby systems
C. For fire pumps
D. Where both single-phase and three-phase power is required

ANSWER _______________

PROCEDURE TO ESTABLISH ANSWER

The question is about the number of services permitted to supply a building or other structure.

1. In Index, find "Services," under which, find "Number, 230.2."

2. Section 230.2(A) and (D) permit additional services for fire pumps, emergency, legally required, and optional standby systems, and different phases.

3. Section 230.2(C)(1) permits additional services for capacity requirements in excess of 2000 amperes but not for 1600 amperes.

4. The correct answer is A.

26

A transformer vault is permitted to be constructed with a 1-hour fire rating where the transformers are protected with any of the following EXCEPT:

A. Water spray
B. Carbon dioxide
C. Halon
D. Foam

ANSWER _______________

PROCEDURE TO ESTABLISH ANSWER

The question is about construction of transformer vaults.

1. In Index, find "Transformer vaults," under which, find "Walls, roofs, and floors, 450.42."
2. Section 450.42 requires the vault to be of 3-hour construction except that a 1-hour rating is permitted where the transformers are protected with automatic sprinkler, water spray, carbon dioxide, or halon. Answers A, B, and C are true. Foam is not included.
3. The correct answer is D.

27

Which one of the following wiring methods is permitted for installing service-entrance conductors?

A. Type NM nonmetallic-sheathed cable
B. Power and control tray cable
C. Type AC armored cable
D. Type MC metal-clad cable

ANSWER _______________

PROCEDURE TO ESTABLISH ANSWER

The question is about wiring methods permitted for service-entrance conductors.

1. In Index, find "Service-entrance conductors," under which, find "Wiring methods, 230.43."
2. Section 230.43 gives 19 types of wiring methods for services 1000 volts or less.
3. Review the methods and find that Type MC cable is the only one of the above methods that is acceptable.
4. The correct answer is D.

28

For grounded systems, a main bonding jumper is the connection between which one of the following?

A. Grounded circuit conductor and equipment grounding conductor at the service
B. Portions of a conductor in a circuit to maintain required ampacity of the circuit
C. Two or more portions of the equipment grounding conductor
D. Grounded circuit conductor and grounding electrode

ANSWER _______________

PROCEDURE TO ESTABLISH ANSWER

The question is about the main bonding jumper.

1. In Index, find "Jumpers, bonding," under which, find "Main," under which, find "Definition, Art. 100."
2. In Article 100, find Bonding Jumper, Main and note the function of main bonding jumper, which is the connection between the grounded circuit conductor and the equipment grounding conductor at the service.
3. The correct answer is A.

29 On a 4-wire delta-connected service where the midpoint of one phase is grounded, the service-entrance conductor having the higher phase voltage to ground shall be durably marked by which one of the following means?

A. Orange in color or by other effective means
B. Red in color only
C. White in color only
D. Any color other than green

ANSWER ______________

PROCEDURE TO ESTABLISH ANSWER

The question is about identifying the high-leg of a 4-wire grounded delta-connected service.

1. In Index, find "Delta-connected," under which, find "Identifying high leg, 3-phase supply, 4-wire, 110.15, 230.56, 408.3(E)(1), 408.3(F), 409.102."
2. Since the question refers to service equipment, find Section 230.56 and find that the color requirement is orange or by other effective means.
3. The correct answer is A.

30 A service is rated 208Y/120-volt, 3-phase. The busbars in the service equipment are rated 800 amperes. The calculated load is 675 amperes. The service disconnecting means consists of two 150-ampere, 3-pole circuit breakers; one 200-ampere, 3-pole circuit breaker; one 400-ampere, 3-pole circuit breaker and one 300-ampere, 3-pole circuit breaker. The equipment is identified as suitable for service equipment. The service-entrance conductors consist of four 500 kcmil Type THW copper conductors in each of two metric designator 78 (Trade Size 3) rigid metal conduits. Is there a violation of the NEC and what would make this installation acceptable, if there is a violation?

A. The service entrance conductors are undersized.
B. The ratings of the disconnecting means do not comply.
C. The permitted conduit conductor fill is exceeded.
D. A single main disconnecting means is required.

ANSWER ______________

PROCEDURE TO ESTABLISH ANSWER

The question is about a service with multiple main breakers serving as the service disconnecting means and the conductors supplying the busing.

1. In Index, find "Service-entrance equipment," sends you to "Service equipment" under which, find "Disconnecting means, under which, find "Six switch rule, 230.71," and "Rating, 230.79, 230.80."

2. There are 5 disconnects described in the question. However, 230.71(B) requires the following: Two to six service disconnects shall be permitted for each service permitted by 230.2 or for each set of service-entrance conductors permitted by 230.40, Exception No. 1, 3, 4, or 5. The two to six service disconnecting means shall be permitted to consist of a combination of any of the following: (1) Separate enclosures with a main service disconnecting means in each enclosure (3) Switchboard(s) where there is only one service disconnect in each separate vertical section where there are barriers separating each vertical section (4) Service disconnects in switchgear or metering centers where each disconnect is located in a separate compartment. Answer D would ensure that the new requirements in 230.71(B) would make this an acceptable installation.

3. Section 230.79 is not applicable; but note Section 230.80, which provides that the combined rating of all switches or circuit-breakers shall be not less than the rating required for a single switch or circuit breaker. The rating required for a single switch would be 700 A. The total of the five disconnects is 1200 A. The total ampacity of the service disconnect ratings is not less than the load. There is no violation of the Code. Answer B does not apply.

4. In Index find "Conductor fill," under which, find "Rigid metal conduit, 344.22." Section 344.22 refers to Table 1, Chapter 9, which in Note 1 refers to Annex C for the maximum number of conductors permitted in the applicable conduit.

5. Table C8 applies to rigid metal conduit and permits four 400 kcmil Type THW conductors in a metric designator 78 (Trade Size 3) conduit. There is no Code violation found. Answer C is not true, as the conduit is not overfilled.

6. In Index, find "Ampacities," under which, find "Conductors, 310.14 through 310.21."

7. Table 310.16 under the column for Type THW 75°C 500 kcmil CU is rated for 380 amperes. 380 amperes x (2) = 760 amperes, which is more than the calculated load of 675 amperes. Therefore, the conductors are properly sized and 230.90(A), Exception No. 2 permits rounding up to an 800-ampere main, and Answer A is not a violation.

   Even though no information was given in the question regarding termination temperatures, 110.14(C)(1)(b) permits the use of the 75°C column as the conductors are larger than 1 AWG.

8. The correct answer is D.

31 Ground-fault protection of equipment (GFPE) shall be provided for electrical services of more than 150 volts to ground, but not exceeding 1000 volts phase-to-phase for each service disconnecting means rated 1000 amperes or more for which one of the following systems?

A. Solidly grounded 3-phase, 4-wire delta systems
B. Open delta systems
C. All grounded systems
D. Solidly grounded wye systems

ANSWER ______________

PROCEDURE TO ESTABLISH ANSWER

The question is about ground-fault protection.

1. In Index, find "Ground-fault protection," under which, find "Service disconnecting means, 230.95."
2. Section 230.95 requires that ground-fault protection of equipment be provided for solidly grounded wye electrical services of more than 150 volts to ground but not exceeding 1000 volts phase-to-phase for each service disconnecting means rated 1000 amperes or more.
3. The correct answer is D.

32 Which one of the following statements regarding cable trays is NOT true?

A. Cable trays are a support method.
B. Unless by exception, cable trays which support service conductors must only contain the service conductors.
C. Cable trays that support other than service conductors are also permitted to support service conductors if provided with a solid fixed barrier to separate the service conductors from the other conductors, and if identified with permanently marked labels.
D. Cable trays are considered by the *NEC* as a raceway.

ANSWER ______________

PROCEDURE TO ESTABLISH ANSWER

The question is about cable trays.

1. In Index, find "Cable Trays, Article 392," under which, find "Uses permitted, 230.44, 392.10."
2. In Sections 230.44 and 392.10, cable trays are described as support methods, not as raceways. Answer A is true; Answer D is not true. Without considering any exception, cable trays which support service conductors are to contain only service conductors. Answer B is true.
3. However, the Exception to 230.44 confirms that Answer C is also true.
4. The correct answer is D.

33 Where ground-fault protection of equipment (GFPE) is required for a solidly grounded wye system the maximum setting of the ground-fault protection of equipment (GFPE) device shall be how many amperes?

A. 300
B. 1000
C. 1200
D. 1600

ANSWER ______________

PROCEDURE TO ESTABLISH ANSWER

The question is about ground-fault settings for ground-fault protection of equipment.

1. In Index, find "Ground-fault protection," under which, find "Service disconnecting means, 230.95."

2. Section 230.95(A) provides that the maximum setting of the ground-fault protection shall be 1200 amperes, etc.

3. The correct answer is C.

34 Barriers shall be placed in all service switchboards such that:

A. They will isolate feeder busbars and terminals from the remainder of the switchboard.
B. Conductors shall be permitted to travel horizontally through vertical sections of switchboards and switchgear where such conductors are isolated from busbars by a barrier
C. They will isolate grounded conductors from equipment grounding conductors.
D. They will isolate the service disconnecting means from the service overcurrent devices.

ANSWER ______________

PROCEDURE TO ESTABLISH ANSWER

The question is about barriers in switchboards.

1. In Index, find "Barriers, 368.234(B)." This only applies to busways and not switchboards. In the index, locate "Switchboards" and then find "Support, busbars, conductors, 408.3."

2. Section 408.3(A)(2), Exception states Conductors shall be permitted to travel horizontally through vertical sections of switchboards and switchgear where such conductors are isolated from busbars by a barrier.

3. The correct answer is B.

35

Conduit or raceways, including their end fittings, shall not rise above the bottom of a switchboard enclosure more than what distance?

A. 75 mm (3 in.)
B. 200 mm (8 in.)
C. 250 mm (10 in.)
D. 300 mm (12 in.)

ANSWER _______________

PROCEDURE TO ESTABLISH ANSWER

The question is about clearances of conduits entering bottoms of switchboards.

1. In Index, find "Switchboards, Art. 408" under which find "Clearances, 110.26, 408.5, 408.18." Sections 110.26 and 408.18 are not applicable to the question as those sections relate to working spaces and clearances from combustible ceilings respectively.

2. Section 408.5 requires that conduit or raceways, including their end fittings, shall not rise more than 75 mm (3 in.) above the bottom of the enclosure.

3. The correct answer is A.

36

An industrial building has an outside transformer of 15 kV rating. The transformer has exposed live parts. A fence enclosing it must be at least what minimum height?

A. 1.8 m (6 ft)
B. 2.1 m (7 ft)
C. 2.5 m (8 ft)
D. 3.0 m (10 ft)

ANSWER _______________

PROCEDURE TO ESTABLISH ANSWER

The question is about enclosing outdoor high voltage equipment or an installation over 1000 volts.

1. In Index, find "Over 1000 volts," under which, find "Enclosure for electrical installations, 110.31, 490-III."

2. Section 110.31 requires a fence of 2.1 m (7 ft) minimum height or features that provide a degree of isolation equivalent to a 2.1 m (7 ft) fence.

3. The correct answer is B.

37

Which one of the following statements about two or more separate sets of service-entrance conductors connected from one service drop or lateral is NOT true?

A. permitted to be installed to each occupancy of a multi-occupancy building.
B. permitted to be installed to a group of occupancies of a multi-occupancy building.
C. permitted to be installed to a single-family dwelling and a detached garage.
D. permitted to be installed to any number of service disconnecting means in separate enclosures grouped at one location on a single-occupancy building.

ANSWER ______________

PROCEDURE TO ESTABLISH ANSWER

The question is about the number of service-entrance conductor sets and where they are permitted.

1. In Index, find "Service-entrance conductors," under which, find "Conductor sets, number of, 230.40."
2. Section 230.40 permits one set of service-entrance conductors for each lateral or drop. There are, however, five exceptions to this general rule.
3. Exception No. 1 permits one set of service-entrance conductors for each class of service installed to each occupancy or group of occupancies in a building with more than one occupancy. Answers A and B are true.
4. Exception No. 2 permits up to six sets of service-entrance conductors to supply up to six separate enclosures. Answer D is not true.
5. Exception No. 3 permits a single-family dwelling and a separate structure, such as a garage, to have one set of service-entrance conductors installed to each. Answer C is true.
6. In the Index, find "Service-entrance equipment," under which, find "Disconnecting means," under which, find "Maximum number of disconnects, six switch rule, 230.71." Section 230.71(A) permits a maximum of six service-disconnect switches in separate enclosures at any one service location, Answer D is not true. There is a limit to the number of service switches allowed.
7. The correct answer is D.

38

What is the minimum vertical distance above grade permitted for the drip loop of service-entrance conductors?

A. 2.5 m (8 ft)
B. 3.0 m (10 ft)
C. 3.7 m (12 ft)
D. 4.5 m (15 ft)

ANSWER ______________

PROCEDURE TO ESTABLISH ANSWER

The question is about the minimum height of drip loops above the ground.

1. In Index, find "Clearances," under which, find "Conductors" under which find "Overhead service, 230.24, 230.26, 551.79." You could also go to "Overhead service conductors" under which find, "Clearances, 230.24"

2. Section 230.24(B)(1) requires the drip loop to be a minimum of 3.0 m (10 ft) above final grade.

3. The correct answer is B.

39

In no case shall the point of attachment of the overhead service conductors to a building be less than what distance above finished grade?

A. 2.5 m (8 ft)
B. 3.0 m (10 ft)
C. 3.7 m (12 ft)
D. 5.5 m (18 ft)

ANSWER ______________

PROCEDURE TO ESTABLISH ANSWER

The question is about the point of attachment of overhead service conductors.

1. In Index, find "Overhead service conductors," under which, find "Point of attachment, 230.26, 230.28."

2. Section 230.26 requires this point to be not less than 3.0 m (10 ft) above finished grade.

3. The correct answer is B.

40

Insulated service-drop or overhead service conductors of the quadruplex type are used for a 480Y/277-volt service to a swimming pool dressing room, pump room, and restroom building. What is the minimum clearance in any direction required between this service drop and a diving platform?

A. 3.0 m (10 ft)
B. 4.4 m (14.5 ft)
C. 5.2 m (17 ft)
D. 5.5 m (18 ft)

ANSWER ______________

PROCEDURE TO ESTABLISH ANSWER

The question is about clearance of service-drop conductors from a swimming pool diving platform.

1. In Index, find "Swimming pools, fountains, and similar installations," under which, find "Overhead conductor clearances, 680.9."

2. Section 680.9(A) refers to Table 680.9(A), which provides clearances between overhead service-drop cables with a solidly grounded bare messenger or neutral conductor, and swimming pools and related equipment. The clearance between the cables in question and the diving platform in any direction (Column B) is 4.4 m (14.5 ft).

3. The correct answer is B.

41

Which one of the following statements about service disconnecting means is NOT true?

A. The service disconnecting means are permitted to be installed in a factory restroom containing only wash basins and toilets.
B. The service disconnecting means shall consist of not more than six switches or circuit breakers.
C. A service disconnecting means for a water pump used for fire protection may be located remote from the other service disconnects.
D. In multiple occupancy buildings, each occupant must have access to his/her service disconnecting means.

ANSWER ______________

PROCEDURE TO ESTABLISH ANSWER

The question is about the installation of service disconnecting means.

1. In Index, find "Service equipment," under which, find "Disconnecting means," under which, find "Location, 230.70(A)," "Multiple occupancy buildings, 230.72(C)," and "Six switch rule, 230.71."

2. Section 230.70(A)(2) states that the service disconnecting means shall not be installed in bathrooms. There is no exception for any occupancy. Answer A is not true.

   Note: According to the definition of bathroom in Article 100, a restroom with only basins and toilets is a bathroom. This definition applies to any occupancy.

3. Section 230.71(A) limits the number of service disconnecting means to six, etc. Answer B is true.

4. Section 230.72(C) requires that each occupant in a multiple-occupancy building shall have access to the occupant's service disconnecting means. Although there is an exception for buildings with continuous building management supervision, no information was given in the question indicating that this is the case. Answer D is true.

5. In Index, find "Fire Pumps." There is no listing for "Disconnecting means" or "disconnecting means location." Therefore, scan Article 695, and find 695.4(B)(3)(a)(4), which requires that fire pump disconnecting means be located sufficiently remote from other buildings or fire pump disconnecting means to prevent it from being inadvertently shut off. Answer C is true.

6. The correct answer is A.

42

Where the service disconnecting means does not disconnect the grounded conductor, other means shall be provided to disconnect the grounded conductor in which one of the following locations?

A. Within the building
B. At any location on the premises
C. Within the service equipment
D. In any panel located within sight of the service equipment

ANSWER ______________

PROCEDURE TO ESTABLISH ANSWER

The question is about the disconnection of the service grounded conductor.

1. In Index, find "Service equipment," under which find "Disconnecting means," under which, find "Grounded conductor, disconnection of, 230.75."

2. Section 230.75, Disconnection of Grounded Conductor, requires the grounded conductor to be disconnected by other means in the service equipment where not disconnected by the service disconnecting means.

3. The correct answer is C.

43

A generator supplying a legally required standby system shall be capable of supplying power to the selected loads within a maximum of how many seconds?

A. 10
B. 30
C. 60
D. 90

ANSWER ______________

PROCEDURE TO ESTABLISH ANSWER

The question is about legally required standby systems.

1. In Index, find "Legally required standby systems," under which, find "Sources of power, 701–III."

2. Scan Article 701, Part III, and find Section 701.12, which provides that current supply is to be available in a time not exceeding 60 seconds.

3. The correct answer is C.

44

Where ground-fault protection of equipment (GFPE) of a service disconnecting means is provided and the available fault current exceeds 10,000 amperes, what is the maximum time-delay setting for the ground-fault protection of equipment device?

A. 0.5 seconds
B. 1.0 seconds
C. 3.0 seconds
D. 5.0 seconds

ANSWER ______________

PROCEDURE TO ESTABLISH ANSWER

The question is about time-delay settings for ground-fault protection equipment for services.

1. In Index, find "Ground-fault protection," under which, find "Service disconnecting means, 230.95."

2. Section 230.95(A) provides that the maximum time delay shall be one second for ground-fault currents equal to or greater than 3000 amperes.

3. The correct answer is B.

45

Service equipment and threaded rigid metal conduit enclosing service-entrance conductors for systems operating at less than 250 volts to ground are permitted to be bonded together by all of the following means EXCEPT:

A. Bonding to the grounded conductor
B. Two standard locknuts, one inside and one outside
C. Bonding-type locknuts and bushings
D. Threaded hubs or bosses

ANSWER ______________

PROCEDURE TO ESTABLISH ANSWER

The question is about bonding of service equipment and service raceways.

1. In Index, find "Bonding," under which, find "Service equipment, 250.92, 250.94."

2. Section 250.92 permits the bonding to be any of the methods specified in answers A, C, and D.

3. Section 250.92(B), first paragraph, states that standard locknuts or bushings shall not be the only means for the bonding required but shall be permitted to be installed to make a mechanical connection of the raceway(s). . Voltage is not a factor when considering the bonding of service equipment.

4. The correct answer is B.

46

What is the minimum size of a copper grounding electrode conductor required for a 600Y/347-volt service for a computer system supplied by two 500 kcmil copper conductors per phase?

A. 2 AWG
B. 1 AWG
C. 1/0 AWG
D. 2/0 AWG

ANSWER ______________

PROCEDURE TO ESTABLISH ANSWER

The question is about the sizing of the grounding electrode conductor.

1. In Index, find "Grounding electrode conductors," under which, find "Sizing, 250.30(A)(6)(a), 250.66, 250.166."

2. Section 250.66, "Size of Alternating-Current Grounding Conductors," refers us to Table 250.66. Table 250.66 requires the size of the grounding electrode conductor to be based on the size of the largest service-entrance conductor or equivalent area for parallel conductors.

3. Two parallel 500 kcmil copper conductors are equivalent to 1000 kcmil. For 1000 kcmil copper, a 2/0 AWG grounding electrode conductor is required.

4. The correct answer is D.

47 Nonshielded cable operating at over 1000 volts may be installed underground in all of the following wiring methods EXCEPT:

A. Rigid metal conduit
B. Intermediate metal conduit
C. Rigid polyvinyl chloride (PVC) conduit encased in 50 mm (2 in.) of concrete
D. Type MC cable where the metallic sheath is grounded through an effective grounding path.

ANSWER ______________

PROCEDURE TO ESTABLISH ANSWER

The question is about wiring methods for underground over 1000-volt installations.

1. In Index, find "Underground wiring," under which, find "Over 1000 volts, 300.50."
2. Section 300.50(A)(2) requires nonshielded cables to be installed in rigid metal conduit, intermediate conduit, or in rigid nonmetallic conduit encased in not less than 75 mm (3 in.) of concrete. Two inches of concrete isn't enough to meet the requirement so Answer C is not permitted encased in 2 inches of concrete.
3. Section 300.50(A)(1) permits Type MC cable to be used if it has a moisture-impervious sheath and meets the other specified conditions.
4. The correct answer is C.

48 Which one of the following statements about splicing direct burial multiconductor cables operating at 7200 volts to ground is NOT true?

A. Underground splices are only permitted if in an approved splice box.
B. Underground splices are required to be watertight.
C. Where cables are shielded the shielding must be continuous across the splice.
D. Underground splices must be protected from mechanical damage.

ANSWER ______________

PROCEDURE TO ESTABLISH ANSWER

The question is about splicing direct burial cables operating at over 1000 volts.

1. In Index, find "Underground wiring" under which, find "Over 1000 volts, 300.50."
2. Section 300.50(D) permits direct burial cables to be spliced without a splice box, etc. Answer A is not true. This section also requires the splice to be watertight and protected from mechanical damage. Answers B and D are true. Where conductors are shielded, the shielding shall be continuous across the splice. Answer C is true.
3. The correct answer is A.

49

Under which of the following conditions is the service disconnecting means permitted to be located remote from the service equipment?

A. Health-care facilities
B. High-rise office buildings
C. Multibuilding, industrial installations
D. None of the above

ANSWER ______________

PROCEDURE TO ESTABLISH ANSWER

The question is about the location permitted or required for the service disconnecting means.

1. In Index find "Service equipment," under which, find "Disconnecting means, 230-VI," under which find "Location 230.70(A)."

2. Scan Section 230.70(A) and find nothing that permits the service disconnecting means to be located remote from the point of entry of the service-entrance conductors. At this point, it is tempting to select choice "D. None of the above." However, we are training you to consider looking at higher voltage installations when you get a question on your test that may appear to be vague. Service equipment can be "1000 Volts or Less" in the Code, or "Over 1000 Volts."

3. Verify whether special consideration might be permitted for over 1000-volt services. In the Index find "Service equipment, 230 Part V" under which find "Disconnecting means" under which find "Over 1000 volts, 230.205, 230.206

4. Section 230.205(C), Remote Control, provides "For multibuilding, industrial installations under single management, the service disconnecting means shall be permitted to be located at a separate building or structure. In such cases, the service disconnecting means shall be permitted to be electrically operated by a readily accessible, remote-control device."

5. The correct answer is C.

50

The minimum size copper grounding electrode conductor for a service operating at 13,800 Y / 7,620 volts supplied by 750 kcmil copper service-entrance conductors, one per phase is:

A. 750 kcmil
B. 3/0 AWG
C. 350 kcmil
D. 2/0 AWG

ANSWER ______________

PROCEDURE TO ESTABLISH ANSWER

The question is about sizing of grounding electrode conductors for systems operating at over 1000 volts

1. In Index, find "Grounding," under which, find "Over 1000 volts, 250–X."

2. In Section 250–Part X, Section 250.180, General, states, "If systems over 1000-volts are grounded, they

shall comply with all applicable requirements of 250.1 through 250.178 and with 250.182 through 250.190, which supplement and modify the preceding sections." Sections 250.182 through 250.190 cover other areas of high-voltage grounding, but not sizing of the grounding electrode conductor specifically.

3. In Index, find "Grounding electrode conductors," under which, find "AC systems, 250.30(A)(6)(a), 250.66, 250.166."

4. Section 250.66, "Size of Alternating-Current Grounding Conductors," refers us to Table 250.66. Table 250.66 requires the size of the grounding electrode conductor to be based on the size of the largest service-entrance conductor or equivalent area for parallel conductors.

5. In Table 250.66, we find the minimum size grounding electrode conductor, which for 750 kcmil copper is size 2/0 copper.

   Note: Section 250.166 states requirements for grounding direct current systems and does not apply here.

6. The correct answer is D.

51 What is the minimum allowed horizontal clearance of overhead service conductors from a storable pool?

A. This is a storable pool, and is not covered by the *NEC*
B. 15 feet
C. 10 feet
D. 18 feet

ANSWER ______________

PROCEDURE TO ESTABLISH ANSWER

The question is about working clearances of conductors from pools.

1. In Index, find "Swimming Pools, fountains, and similar installations," under which, find "Overhead conductor clearances, 680.9"

2. In Section 680.9, 680.9(A) refers us to Table 680.9(A), Overhead conductor clearances. Column C of the table states that the "horizontal limit of clearance measured from the inside wall of the pool" is not to be less than 10 feet.

3. We must also refer to 680.1, Scope, which states that Article 680 applies to permanently installed and to storable (portable) pools.

4. The correct answer is C.

52 Which of the following is NOT a special condition that would allow more than one service to a building or structure?

A. Fire pump service
B. Optional power production services
C. Systems designed for connection to multiple sources of supply for enhanced reliability
D. Emergency, legally required standby, and optional standby power services

ANSWER ______________

PROCEDURE TO ESTABLISH ANSWER

The question is about the number of services permitted to serve a building.

1. In Index, find "Services, Art. 230,"under which, find "Number, 230.2."

2. In 230.2, find "230.2(A), Special Conditions." Answers A, C, and D are true. However, optional power production services are not on the list of services permitted.

3. The correct answer is B.

53 In other than dwelling unit services, service equipment is required to be marked in the field with the available fault current. The field marking shall:

A. Include the name of the individual who applied the marking and the date the marking was applied.
B. Include the date the fault current was calculated and be of sufficient durability to withstand the environment.
C. Include the date the fault current was calculated and the name of the individual performing the calculation.
D. Include the name of the property owner and contact information.

ANSWER ______________

PROCEDURE TO ESTABLISH ANSWER

The question is dealing with the marking the service equipment with the available fault current.

1. There was no direct reference found to the required marking. In the Index find "Marking" 110.21. A review of 110.21 determines it is not applicable to the question; but while scanning the page, find "110.24 Available Fault Current." You can also go to the Index and find "Fault current, available 110.24."

2. The correct answer is B.

54 Equipment such as panelboards and switchboards that are located in other than dwelling units shall be required to be field-marked to warn qualified persons of a potential arc-flash hazard. The marking must be located:

A. so it is visible and not more than 50 ft. from the equipment
B. on the inside of enclosure next to the busing.
C. adjacent to largest branch circuit or feeder overcurrent device
D. so it is clearly visible to the qualified person before examining, servicing or maintaining the equipment.

ANSWER ______________

PROCEDURE TO ESTABLISH ANSWER

The question is related to providing an arc-flash warning to qualified individuals.

1. In the Index find "Warning signs at equipment" under which find "arc-flash hazard warning, 110.16."

2. The correct answer is D.

CHAPTER 8
TRANSFORMERS

1 All of the following statements regarding dry-type transformers are true EXCEPT:

A. Transformers with ventilating openings must be installed so that the ventilating openings are not blocked by walls or other obstructions.
B. The required clearances must be marked on the transformer.
C. Additional heat losses may occur in some transformers, where non-sinusoidal currents are present.
D. Ventilation is not required by the *NEC* for dry-type transformers.

ANSWER _______________

PROCEDURE TO ESTABLISH ANSWER

The question is about general requirements for transformers.

1. In Index, find "Transformers," under which, find "Ventilation, 450.9."
2. In Section 450.9 and Informational Note No. 2, find that answers A, B, and C are all true. Answer D is not true.
3. The correct answer is D.

2 All of the following statements regarding dry-type transformers are true EXCEPT:

A. Appropriate provisions must be made to minimize the possibility of physical damage from external causes to any transformer.
B. The operating voltage of exposed live parts of transformer installations must be indicated by signs or visible markings on the equipment or structures.
C. Only certain energized parts are required to be guarded in accordance with NEC 110.27 and 110.34.
D. Dry-type transformers must be installed in non-combustible moisture-resistant cases or enclosures which prevent the insertion of foreign objects.

ANSWER _______________

PROCEDURE TO ESTABLISH ANSWER

The question is about general requirements for transformers.

1. In Index, find "Transformers," under which, find "Guarding, 450.8."
2. In Section 450.8, find that answers A, B, and D are true. Answer C is not true.
3. The correct answer is C.

3 The full-load amps of a 500 kVA transformer operating at 480Y/277-volts 3-phase is:

A. 500 amperes
B. 60 amperes
C. 601 amperes
D. 1042 amperes

ANSWER _______________

PROCEDURE TO ESTABLISH ANSWER

The question is about calculating the full-load current of a transformer.

1. FLA = kVA x 1000 ÷ volts phase-to-phase x 1.732
2. FLA = 500 x 1000 ÷ 480 volts x 1.732
3. FLA = 500,000 / 831.36
4. FLA = 601 amperes
5. The correct answer is C.

4 What is the maximum permitted fuse-type primary overcurrent protection for a 2000-kVA, 12,470Y/7200-volt, 3-phase, 4-wire transformer with impedance of 5.8 percent?

A. 300 amperes
B. 250 amperes
C. 1600 amperes
D. 1500 amperes

ANSWER _______________

PROCEDURE TO ESTABLISH ANSWER

The question is about sizing overcurrent protection for transformers.

1. In Index, find "Transformers," under which, find "Overcurrent Protection, 450.3, 450.4(A), 450.5(A), 450.5(B)(2), 450.6(B)."
2. In Section 450.3(A), the overcurrent protection for transformers operating at over 1000 volts is to be sized per Table 450.3(A).
3. FLA = kVA x 1000 ÷ volts phase-to-phase x 1.732

   FLA = 2000 kVA x 1000 ÷ 12,470 volts x 1.732

   FLA = 2,000,000 VA ÷ 21,598-Volts

   FLA = 92.6 amperes

92.6 amperes x 300% based on fuse protection = 277.8 amperes.

300% is the maximum amperage per Table 450.3(A) for a transformer operating at over 1000 volts with an impedance of 6% or less.

4. Because 277.8 amperes is not a standard ampere rating for a fuse, the next higher standard rating is permitted to be used per Note 1 to Table 450.3(A).

5. Go to Index. Under "Fuses," find "Ratings, 240.6." In Section 240.6, Table 240.6(A) find that 300 amperes is the next highest standard size.

6. The correct answer is A.

5 What is the maximum permitted circuit-breaker-type, primary-only overcurrent protection for a 50-kVA 480-volt primary and a 120/240-volt, single-phase, 3-wire secondary transformer with an impedance of 5 percent?

A. 130 amperes
B. 80 amperes
C. 125 amperes
D. 600 amperes

ANSWER ______________

PROCEDURE TO ESTABLISH ANSWER

The question is about sizing overcurrent protection for transformers.

1. In Index, find "Transformers," under which, find "Overcurrent Protection, 450.3, 450.4(A), etc."

2. In Section 450.3(B) we find that overcurrent protection for transformers operating at 1000 volts or less is to be sized in accordance with Table 450.3(B).

3. FLA = (kVA x 1000) / (Volts x 1.732)

   FLA = 50,000 VA / (480 x 1.732)

   FLA = 50,000 / 831.36 V

   FLA = 60.14 amperes

   60.14 amperes x 125% = 75.17 amps

   125% is the maximum amperage per Table 450.3(B) for a transformer operating at 1000 volts or less with current of 9 amperes or more. Transformer impedance is not considered here by the *Code* where sizing overcurrent protection for this 50 kVA transformer.

4. Because 75 amperes is not a standard ampere rating for a circuit breaker, the next higher standard rating is permitted to be used in accordance with Note 1 to Table 450.3(A).

5. Go to Index. Under "Circuit breakers and find "Ratings, and "Fixed-trip circuit breakers, 240.6(A) and in Table 240.6(A), find that 80 amperes is the next highest standard size.

6. The correct answer is B.

6

What is the maximum permitted fuse type primary and secondary overcurrent protection for a 75-kVA, 480-volt primary and a 120/240-volt, single-phase, 3-wire secondary transformer with an impedance of 5 percent?

A. 250 amperes, 125 amperes
B. 125 amperes, 250 amperes
C. 375 amperes, 375 amperes
D. 225 amperes, 400 amperes

ANSWER ______________

PROCEDURE TO ESTABLISH ANSWER

The question is about sizing overcurrent protection for transformers.

1. In Index, find "Transformers," under which, find "Overcurrent Protection, 450.3, 450.4(A), etc."

2. In Section 450.3(B) we find that overcurrent protection for transformers operating at 1000 volts or less is to be sized per Table 450.3(B).

3. FLA = (kVA x 1000) / (Volts x 1.732)
FLA = 75,000 VA / (480 x 1.732)
FLA = 75,000 VA / 831.36 V
FLA = 90.21 amperes

   90.21 amperes x 250% = 225.53 amps

   250% is the maximum in accordance with Table 450.3(B) for a transformer primary where primary and secondary protection is provided, operating at 1000 volts or less with current of 9 amperes or more. Transformer impedance is not considered here by the Code where sizing overcurrent protection for this 75-kVA transformer.

4. The calculation is 225.53 amps at 250%, so the primary overcurrent protective device will be 225 amperes .

   Secondary current = _____ FLA
FLA = kVA x 1000 ÷ volts phase-to-phase
FLA = 75 kVA x 1000 ÷ 240 volts
FLA = 75,000 ÷ 240
FLA = 312.5 amperes

   313 amperes x 125% = 391.25 amperes

   125% is the maximum amperage per Table 450.3(B) for a transformer secondary where primary and secondary protection is provided, operating at 1000 volts or less with current of 9 amperes or more. Transformer impedance is not considered here by the *Code* when sizing overcurrent protection for this 75 kVA transformer.

5. Go to Index. Under "Fuses," find "Ratings, 240.6." In Table 240.6(A), find that 375 is the next lower size, and 400 amperes is the next highest standard size.

   Since 391 amperes is not standard ampere ratings for a fuse, the next higher standard rating is permitted to be used for the secondary protection in accordance with Note 1 to Table 450.3(A).

6. The correct answer is D.

PART 2
RANDOM QUESTIONS

Turbocharge your exam preparation by seeing how you do by completing these randomized questions.

[Each question worth 2 points each for a total of 100 points.]

1 What is the minimum ampacity of the phase conductors between the terminals of a 100-ampere rated generator, which is not designed to prevent overloading, to the first overcurrent device?

A. 100
B. 115
C. 125
D. 150

ANSWER ______________

2 On a 4-wire delta-connected service where the midpoint of one phase is grounded, the service-entrance conductor having the higher phase voltage to ground shall be durably marked by which one of the following means?

A. Orange in color or by other effective means
B. Red in color only
C. White in color only
D. Any color other than green

ANSWER ______________

3 Where installed in raceways, conductors larger than what size are required to be stranded?

A. 10 AWG
B. 8 AWG
C. 6 AWG
D. 4 AWG

ANSWER ______________

4 The minimum wire bending space required for deflection of one 750-kcmil cable from the end of the lugs to the wall of the cabinet where the conductors enter or leave the cabinet opposite the terminal is:

A. 203 mm (8 in.)
B. 305 mm (12 in.)
C. 356 mm (14 in.)
D. 432 mm (17 in.)

ANSWER ______________

5 The sizing of conductors supplying several motors is based on which one of the following?

A. Full-load current of all motors in the group
B. Full-load current of all motors plus 25 percent of highest rated motor in group
C. 125 percent of the full-load current of all motors in the group
D. 100 percent of the largest overcurrent device protecting the motors

ANSWER ______________

6 A Schedule 40 rigid polyvinyl chloride (PVC) conduit contains two 500-kcmil Type THW conductors, one 250-kcmil Type THW conductor, and one 1 AWG bare conductor. The minimum permitted conduit size is:

A. Metric Designator 41 (1½ in.)
B. Metric Designator 53 (2 in.)
C. Metric Designator 63 (2½ in.)
D. Metric Designator 78 (3 in.)

ANSWER ______________

7 A single motor air-conditioning unit is rated at 32 amperes. What is the minimum ampacity required for the conductors supplying this unit?

A. 32 amperes
B. 40 amperes
C. 56 amperes
D. 60 amperes

ANSWER ______________

8 Where ground-fault protection of equipment (GFPE) is required for a solidly grounded wye system the maximum setting of the ground-fault protection of equipment (GFPE) device shall be how many amperes?

A. 300
B. 1000
C. 1200
D. 1600

ANSWER ______________

9 Two 500-kcmil conductors per terminal enter or leave the top of a 400-ampere rated panelboard opposite their terminals. No conductors terminate in the bottom of the enclosure. The panelboard shall have a minimum top and bottom wire bending space of:

A. 356 mm (14 in.) of wire bending space at the top and bottom of the panelboard
B. 203 mm (8 in.) of wire bending space at the top and bottom of the panelboard
C. 203 mm (8 in.) of wire bending space at the top and 356 mm (14 in.) of wire bending space at the bottom of the panelboard
D. 356 mm (14 in.) of wire bending space at the top and 203 mm (8 in.) of wire bending space at the bottom of the panelboard

ANSWER ______________

10 The full-load amps of a 500 kVA transformer operating at 480Y/277-volts 3-phase is:

A. 500 amperes
B. 60 amperes
C. 601 amperes
D. 1042 amperes

ANSWER ______________

11 Six 6 AWG THW copper conductors are installed in the same raceway with six 6 AWG Type XHHW-2 copper conductors. All conductors are current carrying. The permitted ampacity for the Type XHHW-2 conductors is:

A. 25.0 amperes
B. 32.5 amperes
C. 38.5 amperes
D. 45.5 amperes

ANSWER ______________

12 Conduit or raceways, including their end fittings, shall not rise above the bottom of a switchboard enclosure more than what distance?

A. 75 mm (3 in.)
B. 200 mm (8 in.)
C. 250 mm (10 in.)
D. 300 mm (12 in.)

ANSWER ______________

13 An outside 480Y/277-volt overhead feeder (open conductors) extends over a commercial area accessible to pedestrians but is not subject to truck traffic. What is the minimum clearance above grade?

A. 2.5 m (8 ft)
B. 3.0 m (10 ft)
C. 3.7 m (12 ft)
D. 4.5 m (15 ft)

ANSWER ______________

14 What is the general lighting load of an office building with an area of 4,500 square feet and the actual number of general-purpose receptacle outlets is unknown?

A. 13,500 VA
B. 15,750 VA
C. 18,000 VA
D. 10,350 VA

ANSWER ______________

15 A generator set has been determined to be vital to the operation of an electrical system. Instead of interrupting the generator circuit, the overload sensing device may:

A. Be omitted
B. Be interlocked
C. Be connected to an annunciator
D. Interrupt the generator field circuit

ANSWER ______________

16 In an existing factory, a 277/480-volt switchboard faces a 120/240-volt panelboard on an opposite wall. If the switchboard is being replaced, and there are qualified persons servicing the installation, and written procedures prevent opening of both enclosures at the same time, what is the minimum working space between the new switchboard and the panelboard?

A. 762 mm (30 in.)
B. 914 mm (36 in.)
C. 1.07 m (3½ ft)
D. 1.2 m (4 ft)

ANSWER ______________

17 Which one of the following taps is NOT permitted to be connected to the supply side of the service disconnecting means?

A. For load management devices
B. For a standby power system
C. To supply outside area lighting
D. To supply fire pump equipment

ANSWER ______________

18 What is the minimum size rigid metal conduit required for two 2/0 AWG and two 1/0 AWG Type THW conductors?

A. Metric Designator 35 (1¼ in.)
B. Metric Designator 41 (1½ in.)
C. Metric Designator 53 (2 in.)
D. Metric Designator 63 (2½ in.)

ANSWER ______________

19 The total inside area of a school is 160,000 square feet, and an outdoor-lighted play area is 120,000 square feet. The total connected load (including the air-conditioning), before applying demand factors is 1,740,000 VA. What is the total VA load for the school, after applying demand factors, as permitted using the optional method of calculation?

A. 1,740,000 VA
B. 1,425,000 VA
C. 480,000 VA
D. 1,305,000 VA

ANSWER ______________

20 Metal cabinets and boxes protected from corrosion solely by enamel are permitted to be used in which one of the following locations?

A. In outdoor locations
B. In car washes
C. In dairies
D. In enclosed parking garages

ANSWER ______________

21 What is the maximum permitted circuit-breaker-type, primary-only overcurrent protection for a 50-kVA 480-volt primary and a 120/240-volt, single-phase, 3-wire secondary transformer with an impedance of 5 percent?

A. 130 amperes
B. 80 amperes
C. 125 amperes
D. 600 amperes

ANSWER ______________

22 A 480-volt, 3-phase feeder of Type THW copper conductors are installed to serve the following five motors: two 25-horsepower, one 20-horsepower, one 7.5-horsepower and one 5-horsepower. The minimum required size of this feeder conductor is:

A. 2 AWG
B. 1 AWG
C. 1/0 AWG
D. 2/0 AWG

ANSWER ______________

23 What maximum percentage shall be used for overload protection to protect motors of one horsepower, marked not over 40°C temperature rise and that start automatically?

A. 100
B. 115
C. 125
D. 140

ANSWER ______________

24 In paralleling conductors of size 1/0 AWG or larger, which one of the following conditions is NOT true?

A. They shall be of the same length and circular mil area.
B. They shall be of the same conductor material and insulation.
C. They shall be of the sunlight resistant type.
D. They shall be terminated in the same manner.

ANSWER ______________

25 A bare copper grounded (neutral) service-entrance conductor is permitted only if the voltage is not over which one of the following?

A. 150 volts to ground
B. 300 volts to ground
C. 600 volts to ground
D. Any voltage in A, B, or C

ANSWER _______________

26 A 480Y/277-volt service is grounded at the utility transformer and enters a building. Under which one of the following conditions is it necessary to bring the grounded conductor into the service equipment?

A. Where only one 480-volt motor is installed
B. Where 3-phase motors and 277-volt lighting is installed
C. Where only 480-volt, 3-phase power is needed
D. All of the above

ANSWER _______________

27 What is the minimum volume required for a junction box containing six 12 AWG and three 10 AWG?

A. 172 cm^3 (10.50 cu. in)
B. 205 cm^3 (12.50 cu. in)
C. 344 cm^3 (21 cu. in)
D. 238 cm^3 (14.50 cu. in)

ANSWER _______________

28 8 AWG Type THHN copper conductors are permitted to be installed in which one of the following locations?

A. Dry locations only
B. Dry and damp locations
C. Wet locations
D. Dry and wet locations

ANSWER _______________

29 What is the minimum working space permitted in front of a panelboard where the voltage to ground is 120 volts?

A. 914 mm (36 in.) wide and 914 mm (36 in.) directly in front
B. 762 mm (30 in.) wide and 914 mm (36 in.) directly in front
C. 762 mm (30 in.) wide and 825 mm (33 in.) directly in front
D. 610 mm (24 in.) wide and 610 mm (24 in.) directly in front

ANSWER _______________

30 An apartment building with a 208Y/120-volt, 3-phase, 4-wire service has 30 units each having 1000 square feet with the following load: electric heat (3 kVA), air conditioner (2 kVA), range (10 kVA), water heater (4 kVA), disposer (0.5 kVA), two vent fans (0.25 kVA) each. There are no laundry facilities. Using the optional calculation method, what is the calculated load for the service equipment?

A. 400 amperes
B. 660 amperes
C. 1300 amperes
D. 1660 amperes

ANSWER _______________

31 What is the minimum size of Type THW copper conductors required to supply a load consisting of a 5-horsepower motor and a 1000-watt resistive appliance on one branch circuit? The circuit operates at 240 volts, single-phase.

A. 10 AWG
B. 8 AWG
C. 6 AWG
D. 1 AWG

ANSWER _______________

32 Which one of the following statements about a cabinet is true?

A. It differs from a cutout box in that the cover is removable.
B. It is intended to be used with switchboards.
C. It cannot be installed in a dwelling unit.
D. It cannot be installed in wet locations.

ANSWER ______________

33 How many 300-watt industrial process lamps may be installed on a branch circuit rated at 15 amperes, 120 volts?

A. 3
B. 4
C. 5
D. 6

ANSWER ______________

34 What is the minimum size Type THW copper conductor required to serve an ac transformer arc welder with a 60 percent duty cycle and at 43 amperes primary current? Note: The termination points (lugs) at the welder are rated at 60°C.

A. 10 AWG
B. 8 AWG
C. 6 AWG
D. 4 AWG

ANSWER ______________

35 Where motor branch-circuit short-circuit and ground-fault protection is provided by time-delay (dual-element) fuses and the rating is not sufficient to start the motor, it shall be permitted to increase the rating to what percentage of the full-load current rating of the motor?

A. 150%
B. 175%
C. 225%
D. 400%

ANSWER ______________

36 What is the minimum wire-bending space required for one 500-kcmil conductor at each lay-in terminal in the panelboard shown in the illustration?

A. 152 mm (6 in.)
B. 279 mm (11 in.)
C. 330 mm (13 in.)
D. 356 mm (14 in.)

ANSWER

37 Which one of the following continuous rigid metal conduit raceway installations between pull points is in compliance?

A. A raceway with three 90-degree and two 45-degree bends
B. A raceway with three 90-degree and three 45-degree bends
C. A raceway with three 90-degree and four 30-degree bends
D. A raceway with two 90-degree, two 45-degree, and four 30-degree bends

ANSWER ______________

38 The controller for a 25-horsepower motor shall have a horsepower-rating not less than what percentage of the horsepower-rating of the motor?

A. 100
B. 115
C. 125
D. 150

ANSWER _______________

39 A panelboard contains 120/240-volt, multi-wire branch-circuits, protected with 30-ampere or less single-pole overcurrent devices. What is the maximum number of overcurrent devices permitted in any one cabinet or cutout box?

A. 18
B. 42
C. 30
D. The maximum number of overcurrent devices for which the panelboard was designed.

ANSWER _______________

40 What is the minimum depth of an outlet box without enclosed devices or utilization equipment?

A. 12.7 mm (½ in.)
B. 23.8 mm (15/16 in.)
C. 31.8 mm (1¼ in.)
D. 38.0 mm (1½ in.)

ANSWER _______________

41 What is the feeder neutral current for a multifamily dwelling supplied with a 120/240-volt, 3-wire service with a calculated load (including demand factors as per Table 220.42) of 111,450 VA, including a 40 kVA range load per Table 220.55?

A. 385 amperes
B. 432 amperes
C. 485 amperes
D. 500 amperes

ANSWER _______________

42 What is the minimum size copper equipment grounding conductor permitted to be used for a circuit having 4/0 AWG phase conductors and protected with 250-ampere fuses?

A. 8 AWG
B. 6 AWG
C. 4 AWG
D. 3 AWG

ANSWER _______________

43 Which one of the following services for an industrial building requires ground-fault protection of equipment (GFPE)?

A. A 208Y/120-volt, 3-phase, 4-wire solidly grounded wye service with a single main disconnect switch rated at 1000 amperes with 800-ampere fuses
B. A 480Y/277-volt, 3-phase, 4-wire solidly grounded wye service with a single main disconnect switch rated at 1000 amperes with 800-ampere fuses
C. A 480-volt, 3-phase, 3-wire service with a 1000-ampere main circuit breaker
D. A 480Y/277-volt, 3-phase, 4-wire solidly grounded delta service with a single main disconnect switch rated at 1000 amperes with 1000 ampere fuses

ANSWER _______________

44 If the disconnecting means for a 230-volt ac, 5-horsepower motor is a motor circuit switch, the motor circuit switch must be horsepower rated not less than what rating?

A. 5 horsepower
B. 7 horsepower
C. 10 horsepower
D. 15 horsepower

ANSWER ______________

45 What is the minimum thickness permitted for a sheet metal steel of an uncoated electrical panelboard cabinet?

A. 1.35 mm (0.053 in.)
B. 1.52 mm (0.060 in.)
C. 1.59 mm (0.063 in.)
D. 2.29 mm (0.090 in.)

ANSWER ______________

46 Conductors supplying the following group of 3-phase, 230-volt induction motors: one 3-horsepower, two 5-horsepower and one 7½-horsepower, shall have a minimum ampacity of how many amperes?

A. 62.0
B. 65.3
C. 67.5
D. 77.5

ANSWER ______________

47 The ampacity of a 2-wire, 18 AWG Type SO flexible cord is:

A. 7 amperes
B. 10 amperes
C. 15 amperes
D. 20 amperes

ANSWER ______________

48 What is the minimum headroom required for the working space in front of an 800-ampere service-disconnect switch?

A. 1.8 m (6 ft)
B. 2.6 m (6 ft 3 in.)
C. 2.0 m (6 ft 6 in.)
D. 2.5 m (8 ft)

ANSWER ______________

49 The minimum number of overload units providing overload protection for a 3-phase ac motor connected to a 208Y/120-volt, 3-phase, 4-wire system is:

A. 1
B. 2
C. 3
D. 4

ANSWER ______________

50 For branch-circuit or feeder calculations in industrial occupancies, what is the unit load for each outlet supplying heavy-duty lampholders that are not used for general illumination?

A. 180 volt-amperes
B. 600 volt-amperes
C. 20 amperes
D. 5 amperes

ANSWER ______________

PART 2
ANSWERS

Grade yourself against the previous questions.

1.	**B**	Motors, Question 29
2.	**A**	Service Equipment, Question 29
3.	**A**	Conductors, Question 28
4.	**D**	Cabinets, Question 24
5.	**B**	Conductors, Question 22
6.	**D**	Calculations, Question 45
7.	**B**	Motors, Question 11
8.	**C**	Service Equipment, Question 33
9.	**D**	Cabinets, Question 29
10.	**C**	Transformers, Question 3
11.	**B**	Conductors, Question 31
12.	**A**	Service Equipment, Question 35
13.	**C**	Conductors, Question 19
14.	**D**	Calculations, Question 30
15.	**C**	Motor, Question 34
16.	**C**	Cabinets, Question 31
17.	**C**	Service Equipment, Question 14
18.	**C**	Conductors, Question 26
19.	**B**	Calculations, Question 9
20.	**D**	Cabinets, Question 20
21.	**B**	Transformer, Question 5
22.	**B**	Conductors, Question 4
23.	**C**	Motors, Question 50
24.	**C**	Conductors, Question 35
25.	**D**	Service Equipment, Question 12
26.	**D**	Service Equipment, Question 2
27.	**C**	Calculations, Question 65
28.	**B**	Conductors, Question 14
29.	**B**	Cabinets, Question 32
30.	**B**	Service Equipment, Question 8
31.	**B**	Motors, Question 38
32.	**A**	Cabinets, Question 17
33.	**B**	Calculations, Question 18
34.	**B**	Calculations, Question 42
35.	**C**	Motors, Question 42
36.	**B**	Cabinets, Question 8
37.	**A**	Calculations, Question 27
38.	**A**	Motors, Question 59
39.	**D**	Cabinets, Question 1
40.	**A**	Cabinets, Question 33
41.	**A**	Calculations, Question 34
42.	**C**	Conductors, Question 9
43.	**B**	Service Equipment, Question 20
44.	**A**	Motors, Question 20
45.	**B**	Motors, Question 39
46.	**C**	Calculations, Question 1
47.	**B**	Calculations, Question 24
48.	**C**	Service Equipment, Question 24
49.	**C**	Motors, Question 54
50.	**B**	Calculations, Question 15

NOTES

PLAN REVIEW STUDY GUIDE

ELECTRICAL INSPECTOR CERTIFICATION PROGRAM

President/CEO
Rudolph Garza

Director of Digital Education / Technical Editor
Joseph Wages, Jr.

Technical Review
Mark Ode
Joseph Wages, Jr.

Cover Design / Desktop Publishing
Laura Hildreth

Composed at:
International Association of Electrical Inspectors
in Prohibition and Adobe Garamond Pro by Adobe©

Made in the USA
Monee, IL
03 June 2023

35230723R00116